F

Third Edition

D1445018

Ronald E. Ney, Jr. Ph.D.

 Government Institutes, Inc., Rockville, MD

Government Institutes, Inc., 4 Research Place, Rockville, Maryland 20850, USA.
Phone: (301) 921-2300
Fax: (301) 921-0373
Email: giinfo@govinst.com
Internet address: http://www.govinst.com

02 01 00 99 98   5   4   3   2   1

Chapters 1–5 of this book were originally published as *Where Did That Chemical Go?* (New York: Van Nostrand Reinhold, 1990).

Library of Congress Cataloging-in-Publication Data

Fate and transport of organic chemicals in the environment : a practical guide/ by Ronald E. Ney.--3rd ed.
     p. cm.
   Includes index.
   ISBN: 0-86587-626-6
   1.Organic compounds--Environmental impacts. 2. Pollution--Risk assessment.
     I. Ney, Ronald E. 1936– Where did that chemical go? II. Title.
  TD196.073N49   1995
  628.5 '2--dc20

                95-31004
                CIP

Printed in the United States of America

Contents

Preface / v
Acknowledgements / vii
Chapter 1. Fundamentals / 1
    Fate and Transport / 1
    Environmental Compartments / 4
Chapter 2. Physical and Chemical Processes / 9
    Water Solubility / 9
    Octanol Water / 12
    Hydrolysis / 14
    Photolysis / 17
    Volatilization / 18
    Soil Sorption / 19
    Leaching in Soil / 22
    Runoff / 24
Chapter 3. Biological Processes / 25
    Biodegradation / 25
    Bioaccumulation / 26
Chapter 4. Exposure Assessment / 29
Chapter 5. Examples / 31
    Chemicals / 32
Chapter 6. Predictive Methods for Pesticides / 173
Chapter 7. Effects of the Environment on Chemicals / 177
    Chemical Mixtures / 178
    Hazardous Organics / 179
    Immobilization / 180
    Prevention of Contamination / 180
    Higher Molecular Weights Compounds / 181
    Studies Needed for Assessments / 181
    Type Laboratory Test Needed / 181

Chapter 8. Leaching in Soil / 185
Chapter 9. Octanol-Water Partition Coefficient / 195
Chapter 10. Water Solubility / 205
Chapter 11. Sorption In Soil / 217
Chapter 12. Bioconcentration and Bioaccumulation / 227
Chapter 13. Pesticide Use Predictions / 243
Chapter 14. Similar Chemical Structures / 255
        Aldehyde / 255
        Anilide, Amine, Amino, and Amide / 256
        Aniline / 257
        Azo and Nitrosoaniline / 258
        Benzene / 258
        Imidazole / 258
        Biphenyl/Diphenyl / 259
        Thiocarbamate / 259
        Imide and Imine / 260
        Indole / 260
        Amido / 261
        Phthalimide / 261
        Chloroaniline / 261
        Vinyl / 262
        Hydrazine / 262
        Steroid Hormones / 264
        Safroles / 265
        Chrysenes / 265
Chapter 15. Dissipation in Soil / 267
        Scenario 1 / 268
        Scenario 2 / 268
        Scenario 3 / 268
        Scenario 4 / 269
Chapter 16. Estimated Exposure to Fish—Safety Factor (SF) / 271
Chapter 17. Chemical Exposure In and Out Door / 277
        Are You Poisoning Yourself? / 277
        Realty / 280
Chapter 18. Predictive Equations for Organic Chemical
        Compounds / 287
Glossary / 361
Index / 363
About the Author / 371

Preface

Would you like to be able to predict potential exposure to a toxic chemical? Of course you would. Would you like to be able to prevent your exposure to a toxic chemical. Of course you would, if possible. This book will help you predict and prevent your exposure to toxic chemicals through tracking the fate and transport of chemicals in the environment.

Chapter 1 demonstrates what the book is about the use of techniques to predict the fate and transport of a chemical in the environment. Definitions are given in terms simple enough for any reader of the book to understand them. Chapters 2 and 3 present predictive techniques and ways of using them. Tables are given in which users can insert data to make the predictions. Chapter 4 ties together the material of the previous chapters, showing how data and predictive techniques may be combined to assess the route of exposure. Chapter 5 gives examples of specific chemicals, showing different types of data, and guides the user, as if hands on, through a step-by-step prediction procedure. The reader will learn how to use and interpret pertinent data and how to predict the fate and transport of a chemical, as well as to predict exposure routes.

Chapter 6 begins to focus on pesticides. This half of the book, which comprises the second edition, presents predictive methods and provides a unique tool to help you predict the potential for exposure to toxic chemicals, mainly pesticides.

Chapter 7 presents general discussions on chemicals in different environments so that the reader will be cognizant of his or her environment and the environment of others.

Chapters 8 through 12 discuss mechanisms of prediction, techniques for using the mechanisms, and practice in making assessments on potential exposure. Exercises and answers are also given to help the reader practice using these mechanisms.

Chapter 13 discusses pesticide dosage and safety predictions moving towards acceptable pesticide use patterns.

Chapter 14 discusses the use of chemical structure to predict adverse effects.

Chapter 15 discusses dissipation in soil.

Chapter 16 presents a hypothetical method to prevent environmental aquatic kills.

Chapter 17 presents discussions on chemicals in different environments so the reader will be cognizant of his or her environment and the environments of others.

Chapter 18 presents equations for predicting the the fate and transport of chemicals other than pesticides, although pesticides fit into some of the presented equations.

The entire book will enable you to assess whether a chemical will bioaccumulate in animals. If bioaccumulation does occur, there is a good possibility that the chemical can be passed on the future generations through sperm, ovum, blood, and milk.

The book is intended for use by either laypersons or scientists. It will enable real estate agents and appraisers, lending institutes, developers, waste managers, chemical manufacturers and users, environmentalists, educators, government officials, environmental assessors, private citizens, and others to predict potential routes of exposure to chemicals.

As you read the book, I hope you will learn the fate and transport of organic chemicals in the environment and become able to predict your potential exposure to a toxic chemical.

Acknowledgments

I am grateful for the support of my wife, Sue McClure Ney, who provided not only her word processing skills but generous portions of encouragement, consideration, and love as well, during the preparation of this book. I also thank Rene Ramos for contributing the illustration of environmental compartments.

FATE
and
TRANSPORT
of
ORGANIC
CHEMICALS

in the

ENVIRONMENT

A Practical Guide

Chapter 1
Fundamentals

FATE AND TRANSPORT

Many questions must be considered when we attempt to determine what happens to a chemical and its breakdown products (i.e., degradates, biodegradates, metabolites, transformation products, dissociation products, hydrolytic products) in the environment. After we obtain data about such chemicals, it can be quite difficult to interpret them. The intention of this book is to present an overview of the data that are obtainable and of their interpretation.

We will: (1) discuss a chemical as the parent chemical because almost any organic chemical, parent or breakdown product, will fit the examples given; (2) present discussions on what could happen to a chemical in an environmental compartment; (3) show how to make predictions on chemical fate in the environment; and (4) demonstrate how, in most cases, to determine and predict the fate of an organic chemical in a particular environmental compartment when few or no data are available. The overall purpose of this book is that it should offer insight and understanding, as well as be a guide to predicting the fate and transport of chemicals (mostly organic chemicals) in the environment.

There are many questions to answer in the assessment of the fate of a chemical in environmental compartments. Six are considered here.

1. *What are the mechanisms that break down a chemical to a toxic, nontoxic, or naturally occurring chemical(s)* [1]? These mechanisms include the following:

   a. Photolysis. If a chemical can sorb sunlight, the chemical can be transformed in media (i.e., air, water, and soil) or on media (i.e., water, soil, plants, and animals) to produce transformation products.

   b. Biodegradation in all environmental compartments to produce biodegradates, if microbes are present.
   c. Metabolism in plants and animals to produce metabolites.
   d. Hydrolysis in all environmental compartments to produce hydrolytic products.
   e. Dissociation in all environmental compartments to produce dissociated products.
   f. Sorption in all environmental compartments.
   g. Bioaccumulation in plants and animals.

2. *What is the rate of dissipation or half-life ($t_{1/2}$) of a chemical in environmental compartments* [1]? The rate of dissipation or the half-life is the time that it takes for the chemical to be reduced by one-half of its original amount, measured from the time of its introduction into an environmental compartment (i.e., when 50%) of the chemical has vanished and only 50% of it remains. The lost 50% could be in the form of a breakdown product(s), either toxic, nontoxic, or naturally occurring. This is why in real situations breakdown products must be identified and studied in the same way that the parent chemical is studied. The dissipation of a chemical may be real, resulting in the breakdown of the parent chemical (usually organic chemicals), or not real, due to adsorption or movement of the chemical from contaminated areas to noncontaminated areas. There are ways to predict this, even when no or few data are available, as will be discussed later. In any event, these aspects should be determined by sorption or mobility studies. One must always remember that the fact that a chemical cannot be found does not mean it is not there. Good testing will establish its whereabouts.

3. *How is a chemical moved in the environment by natural means* [1]? These means include:

   a. Volatilization: A chemical that is volatile can move into the air; and once in the air, the chemical can be transformed, hydrolyzed, dissociated, biodegraded, or sorbed to dust particles. The chemical also can be released from air to contaminate soil, water, plants, and animals via dust particle fallout or by precipitation.
   b. Leaching: A chemical may be transported through soil by solvents (water, etc.) or with soil movement. Any chemical that leaches can contaminate groundwater, an action that in turn may result in the contamination of air, soil, other waters, plants, and animals.

c. Runoff: A chemical may move across a surface with a solvent or with soil movement to contaminate air, water, soil, plants, and animals.

d. Food-chain contamination: A chemical that is passed from one environment can be passed on to water, air, soil, plants, or animals and on up the food chain. For example:

   (1) A chemical that gets into the aquatic environment can be taken up by microscopic organisms, which are eaten by fish, which are eaten by wildlife and humans.

   (2) A chemical in water could reach crops via irrigation, and the crops could be eaten by wildlife and humans.

4. *What happens to a chemical adsorbed or absorbed in soil or in plants or animals* [1]? We will only discuss sorption in soil; because sorption in plants and animals is beyond the scope of this book although it will be discussed in general, in terms of bioaccumulation. These are the possibilities in soil:

a. Absorption: A chemical that is absorbed in soil can be released in any environment (like water being released from a sponge).

b. Adsorption: A chemical that is adsorbed to soil particles is unlikely to be leached by water. An adsorbed chemical may be leached with other solvents (if they get into soil), by soil movement, or by water if soil sorption sites have been filled or taken up by another chemical. Another cause of leaching is saturation; that is, leaching occurs as the starting chemical is continuously added to the soil, thus saturating the soil sites. (Consider the analogy of a parking lot that is full; where do the cars park?) A plant can release soil-bound residues via its root system and translocate the chemical throughout the plant. Likewise, animals that ingest soil can release the chemical residues adsorbed in the soil and translocate them throughout their bodies. These residues could be absorbed in the fat tissue or by protein in the animals [2].

5. *How does food-chain contamination occur* [1]? This type of contamination occurs when one environment is contaminated by a chemical, and that chemical is released into another environment, the end result being that animals ingest the chemical. For example:

a. A chemical that gets into air can be released onto soil, water, plants, and animals, and then noncontaminated animals may eat, drink, or breathe the contamination.

b. Fish may bioaccumulate the chemical in their tissue, and when other animals eat the fish, they in turn also will become contaminated. Thus food-chain contamination has occurred, followed by bioaccumulation up the food-chain ladder.

6. *What are accumulation and bioaccumulation?* These terms refer to the condition in which a chemical builds up and does not dissipate; in soil it is called accumulation, and in plants or animals it is called bioaccumulation. The chemical is not broken down in an environmental compartment. In mammals another problem exists—the potential for residues to be passed on into mothers' milk, then to nursing infants.

REFERENCES

[1] Ney, Ronald E., Jr., "Regulatory Aspects of Bound Residues," Workshop V-C, 4th International Congress of Pesticide Chemistry, International Union of Pure and Applied Chemistry, Zurich, Switzerland, Unpublished paper, 1978.

[2] Yip, George and Ronald E. Ney, Jr., "Analysis of 2,4-D Residues in Milk and Forage," *Weeds, Journal of the Weed Society of America,* vol. 14, pp. 167-170, April 1966.

ENVIRONMENTAL COMPARTMENTS

Five environmental compartments are considered herein. Many chemical, physical, and living reactions can occur in these compartments; and before any data can be discussed, we must know what the compartments are, and what could occur in each one. We now consider relevant features of each compartment.

Compartment Air

- *Contamination:* Air contamination occurs when a volatile chemical or airborne particulate matter (dust) containing a chemical gets into the air as a result of a spill, evaporation, or any release.
- *Reactions:* A chemical can be phototransformed in air, or it can be sorbed to particulate matter and be biodegraded, dissociated, hydrolyzed, or phototransformed.
- *Mobility:* Chemicals can be moved throughout air, by air or precipitation, or can move as fallout with precipitation or with particulate matter to contaminate other environmental compartments.
- *Exposure:* Airborne chemicals could result in chemical exposure of all environmental compartments [1].

Compartment Water

- *Contamination:* Water contamination occurs by fallout from air, from spills, from substances directly applied or intentionally put into water, with runoff, or from leaching into water.
- *Reactions:* Depending on the water (i.e., streams, lakes, ponds, groundwater, ocean, etc.), reactions may include dissociation, hydrolysis, phototransformation, biodegradation, or sorption to particulate matter.
- *Mobility:* Movement may occur with volatilization, water movement, evaporation, irrigation with well water, or animals, resulting in environmental contamination.
- *Exposure:* Chemicals in this compartment could result in the chemical exposure of all environmental compartments [l].

Compartment Soil

- *Contamination:* Soil contamination occurs by spills, fallout from air, or substances directly or indirectly applied to or put into or on soils.
- *Reactions:* Hydrolysis, dissociation, sorption, biodegradation, and photolysis reactions are possible.
- *Mobility:* Volatilization, runoff, leaching, or plant and animal uptake resulting in food-chain contamination may occur.

- *Exposure:* Soil contamination could result in the chemical exposure of all environmental compartments [1].

Compartment Plants

- *Contamination:* Plant contamination may result from fallout, spills, substances indirectly or directly applied to soils, irrigation, and materials in manure and in compost.

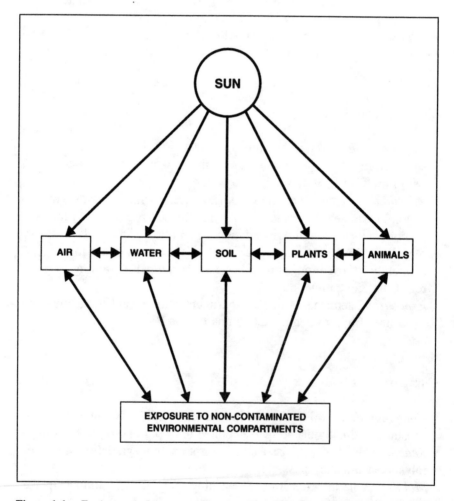

Figure 1-1. Environmental compartments. A never ending circle, from self-contamination to contamination of noncontaminated compartments and back again.

- *Reactions:* Metabolism, hydrolysis, dissociation, photolysis on the plant surface, or sorption may occur.
- *Mobility:* Movement may be occur by release into the air, into the soil via the root system, into the air if the plant is burned, or into the food-chain if the plant is eaten.
- *Exposure:* Plant contamination could result in the chemical exposure of all environments [1].

Compartment Animals

- *Contamination:* Animal contamination may occur by fallout, substances directly or indirectly applied to or on animals, eating other plants or animals, drinking water, or breathing.
- *Reactions:* Metabolism, hydrolysis, dissociation, bioaccumulation, and phototransformation on surfaces are possible.
- *Mobility:* Respiration, excretion, and release if the substance is burned or eaten are means of movement.
- *Exposure:* Animal contamination could result in exposure of all environmental compartments [1].

Thus we have seen: how environmental compartments are contaminated; reactions that could occur to chemicals in those compartments; mobility in and out of the compartments, and how one compartment can contaminate other noncontaminated compartments; and the exposure possibilities (see Figure 1-1 and Table 1-1).

Table 1-1. Reactions in Environmental Compartments.

REACTIONS	AIR	WATER	SOIL	PLANTS	ANIMALS
Hydrolysis	X	X	X	X	X
Phototransformation	X	X	X	X	X
Dissociation	X	X	X	X	X
Solubility	X	X	X	X	X
Sorption	X	X	X	X	X
Biodegradation	X	X	X	X	X
Metabolism				X	X
Accumulation			X		
Bioaccumulation				X	X
Volatilization	X	X	X		
Respiration				X	X
Excretion					X

REFERENCES

[1] Ney, Ronald E., Jr., "Exposure Assessment Considerations and Problems," Exposure Assessment Workshop, U.S. Environmental Protection Agency, UP, 1982.

Chapter 2
Physical and Chemical Processes

WATER SOLUBILITY

Water solubility is perhaps the most important chemical characteristic, used to assess (1) chemical mobility, (2) chemical stability or breakdown, (3) chemical accumulation, (4) chemical bioaccumulation, and (5) chemical sorption in any environmental compartment. Water solubility should be one of the easiest chemical test methods; however, if you were to look up the solubility of DDT, you would find a long list of solubilities. Luckily those values are presented in a range that enables one to make some kind of assessment or prediction; and that is the purpose of our discussions.

What can water solubility be used for? Remember these key points:

1. The higher the water solubility of a chemical, the more likely it is to be mobile, and the less likely it is to be accumulative, bioaccumulative, volatile, and persistent; and a highly soluble chemical is prone to biodegradation and metabolism that may detoxify the parent chemical.
2. The lower the water solubility of a chemical, the more likely it is that it will be immobilized via adsorption, and thus it is less mobile, more accumulative or bioaccumulative, persistent in environmental compartments, and slightly prone to biodegradation, and it may be metabolized in plants and animals.
3. The in-between range of high and low water solubilities indicates chemicals whose behavior could go either way, as discussed in 1 and 2 above.

The above generalities are meaningless unless we have values or ranges on which to base predictions. Thus, to discern the fate of chemicals by water solubilities, let us assign the following numerical values, presented by Ney [3]:

1. Low solubility: less than 10 ppm (<10 ppm).
2. Medium solubility: between 10 and 1 ,000 ppm.
3. High solubility: greater than 1,000 ppm (>1,000 ppm).

Table 2-1 may be used to link the environmental fate of chemicals to their water solubility (WS).

The relationships are not clear-cut, as the water solubility of a chemical can also affect other degradative or transformation process, such as hydrolysis, photolysis, and/or biodegradation. Water solubility can be used to predict sorption or desorption with soil organic matter/organic carbon, mobility/leaching, and bioaccumulation in animals by means of lipo-solubility prediction. These aspects of its use will be discussed in other chapters.

Proven, validated test methods should be used to measure the water solubility of a chemical. If at all possible, a radiolabeled parent chemical should be used. The use of a radiolabeled chemical will enable the researcher to discern whether sorption to equipment, phototransformation (easily prevented), or dissociation has occurred, and, most important, to validate chemical stability, solubility, and analytical methodology. A radiolabeled chemical should be used to validate most studies.

In the environment there would be many effects on chemical solubility, such as the effects of temperature, the concentration of chemicals, sorption, and so on.

Table 2-1. Water Solubility (WS)

ENVIRONMENTAL COMPARTMENTS	LOW < 10PPM	MEDIUM 10-1,000PPM	HIGH > 1,000PPM
Mobility	n *	either way	yes
Adsorption	yes	either way	n
Biodegradation	maybe	either way	yes
Metabolism	maybe	either way	yes
Accumulation	yes	either way	n
Bioaccumulation	yes	either way	n
Persistence	yes	either way	n
Dissipation	n to slowly	either way	yes
Food-chain contamination	yes	either way	n

* n denotes negligible.

If no data on solubility exist, then one could predict solubility by using chemical structure methodology, an advanced predictive technique for chemists. I have found the best illustrations in the *Handbook of Chemical Property Estimation Methods* by Warren J. Lyman, William F. Reehl, and David H. Rosenblatt, published by the McGraw-Hill Book Company, a source that will be cited frequently in this book [2]. Kenaga [1] also has reported on mathematical equations used to calculate solubility.

Lyman et al. [2] reported dichloro diphenyl trichloroethane (DDT) to have a measured solubility of 1.2 µg/L or 1.7 µg/L, indicating a large range for error (200%). Lyman et al. [2] also reported a water solubility, using the chemical structure of DDT, of 3.89 µg/L. Other sources report that DDT is almost insoluble. For the purpose of this discussion, let us use the solubilities of 1.2 µg/L and 3.89 µg/L to predict what may happen in the environment. To do this, we will convert µg/L to ppm, for use with Table 2-1. The following example will help you to understand how to convert µg/L (ppb) to mg/L (ppm) (in this case, 1.2 µg/L):

ppm = mg/L (parts per million = milligrams per liter)
ppb = µg/L (parts per billion = micrograms per liter)
µg/L = one millionth of a gram per liter
mg/L = one thousandth of a gram per liter

$$\frac{1{,}000 \text{ mg/L}}{1{,}000{,}000 \text{ µg/L}} \times 1.2 \text{ µg/L} =$$

$$\frac{0.001 \text{ mg/L} \times 1.2 \text{ µg/L}}{\text{µg/L}} = 0.0012 \text{ mg/L (ppm)}$$

DDT has a water solubility in the range of 0.0012 ppm to 0.00389 ppm; thus, it falls at the level of <10 ppm (see Table 2-1).

With this one piece of data, we can say that DDT will persist in any environmental compartment, accumulate in soil, bioaccumulate in plants and animals, and bioconcentrate in the food-chain. The same data can be used in other predictive methods, which will be discussed in following chapters.

If the solubility of a chemical were 1,000 ppm, then predictions about its fate could be based on Table 2-1. If it were in the range of 10 to 1,000 ppm, problems could arise, and additional data might be needed to predict its fate in environmental compartments. The use of other predictive methods will be discussed in following chapters.

REFERENCES

[1] Kenaga, E. E. and C. A. I. Goring, "Relationship Between Water Solubility, Soil Sorption, Octanol-Water Partitioning, and Concentration of Chemicals in Biota," Special Technical Publication 707, American Society for Testing and Materials, 78-115, 1980.
[2] Lyman, Warren J., William F. Reehl, and David H. Rosenblatt, *Handbook of Chemical Property Estimation Methods,* McGraw-Hill Book Company, 2-44 and 2-45, 1982.
[3] Ney, Ronald E., Jr., "Fate, Transport and Prediction Model Application to Environmental Pollutants," Spring Research Symposium, James Madison University, Harrisonburg, Virginia, UP, April 16,1981.

OCTANOL WATER

The octanol water partition coefficient (Kow), or partition coefficient (P), is an indicator of the bioaccumulation or bioconcentration potential of a chemical in the fatty tissue of living organisms. The Kow or P value (which has no units) is an indicator of water solubility, mobility, sorption, and bioaccumulation. The symbols Kow and P are used interchangeably, but this book will use only Kow.

The Kow represents a mathematical equation expressing the ratio of the equilibrium concentrations of a chemical in octanol and water phases; that is, it is the ratio of the concentration of a chemical in octanol to the concentration of that chemical in water at equilibrium. Simply put, it is the ratio of an organic chemical's distribution between octanol and water phases:

$$\text{Kow} = \frac{\text{concentration of organic chemical in octanol phase}}{\text{concentration of organic chemical in water phase}}$$

The use of the Kow depends on its size:

1. The higher the Kow is, the greater the affinity of the chemical to bioaccumulate/bioconcentrate in the food chain, the greater its potential for sorption in soil, and the lower its mobility. This also means lower solubility in water. Do not confuse water solubility with Kow, as the ratio does not express water solubility, and there are no units of measure, just a number.

2. The lower the Kow is the less the chemical's affinity to bioaccumulate, the greater its potential for mobility, the greater its solubility, and the greater its potential to biodegrade and to be metabolized by plants and animals.

Here also we need numbers for assessment of chemical fate; thus, I have made the following numerical assignments [3]:

1. A Kow of less than 500 (<500) would be indicative of high water solubility, mobility, little to no bioaccumulation or accumulation, and degradability by microbes, plants, and animals.
2. A high Kow, greater than 1,000 (>1,000), is indicative of low water solubility, immobility, nonbiodegradability, and a chemical that is bioaccumulative, accumulative, persistent, and sorbed in soil.
3. A midrange Kow, 500 to 1,000, indicates that the chemical can go the way of either low or high Kow.

Table 2-2 show how to predict the environmental fate of chemicals by using the Kow.

There are many ways to obtain a Kow, by either laboratory test or mathematical equations. Proven and validated test methods should be chosen. If at all possible, a radiolabeled parent chemical should be used; the reasons for doing so were discussed in the section on water solubility.

If no data exist, chemical structure can be used to predict the Kow, [2]. Lyman et al. [2] and Kenaga [l] have reported on the use of mathematical equations for Kow prediction.

Table 2-2. Octanol Water Partition Coefficient and Fate of Chemicals

ENVIRONMENTAL COMPARTMENTS	LOW KOW <500	MEDIUM KOW 500–1,000	HIGH KOW >1,000
Persistent	n *	either way	yes
Adsorbed	n	either way	yes
Absorbed	yes	either way	n
Biodegraded	yes	either way	n to slowly
Metabolized	yes	either way	n to slowly
Accumulated	n	either way	yes
Bioaccumulated	n	either way	yes
Dissipated	yes	either way	n

* n denotes negligible.

Kenaga [1] reported DDT to have a Kow of 960,000. Using Table 2-2, one can easily see that a Kow of 960,000 is > 1 ,000, indicating that DDT has low water solubility, is persistent, is adsorbed in soil, is slowly biodegraded, is slowly metabolized, and is accumulated and bioaccumulated, so that food-chain contamination can be predicted. These are a different expression of the predictions based on the water solubility of DDT.

REFERENCES

[1] Kenaga, E. E. and C. A. I. Goring, "Relationship between Water Solubility, Soil Sorption, Octanol-Water Partitioning, and Concentration of Chemicals in Biota," Special Technical Publication 707, American Society for Testing and Materials, 78-115, 1980.
[2] Lyman, Warren J., William F. Reehl, and David H. Rosenblatt, *Handbook of Chemical Property Estimation Methods,* McGraw Hill Book Company, Chapter 1, 1982.
[3] Ney, Ronald E., Jr., "Fate, Transport and Prediction Model Application to Environmental Pollutants," Spring Research Symposium, James Madison University, UP, April 16, 1981.

HYDROLYSIS

Hydrolysis is perhaps one of the most important mechanisms in the environment for the breakdown of a parent chemical [2]. It occurs in soils, water, plants, animals, and possibly air because water exists in all of these environments. The hydrolysis of pesticides has even occurred on plant surfaces.

Many environmental factors influence the rate of hydrolytic degradation, such as temperature, pH, solubility, sunlight, ad- or absorption, volatility, and so on. The rate of hydrolization of a chemical is the time that it takes to reach one-half ($t_{1/2}$) of its original amount. Because many outside influences can affect the rate of hydrolysis, study of the process should be done under controlled conditions using a radiolabeled parent molecule. The use, if applicable, of a radiolabeled chemical enables the researcher to study sorption, breakdown products, and recovery (to validate analytical methods).

There are chemical structure prediction methods and mathematical equations available to predict hydrolysis, as reported by Lyman et al. [1], but these approaches are for the experienced professional.

The rate of hydrolysis must be known to determine persistence in the environment. (Persistence is how long a parent chemical will be present for exposure.) There are no clear-cut assessments for half-lives because exposures differ in the environmental compartments. The rate of hydrolysis can be used in exposure assessments for toxic and phytotoxic chemicals, but toxicity considerations are not discussed herein, as they are beyond the scope of this book.

Water Compartments

No half-life is acceptable if the chemical causes immediate harm to organic organisms, or harm if immediately used for irrigation or potable water. If there are no immediate concerns, then a half-life of 30 days may be acceptable if no accumulation/bioaccumulation occurs. Because water cannot be controlled, care has to be taken in assessing exposure. It would be best to keep contaminating chemicals out of water.

Soil Compartments

In most cases soils can be controlled; thus the half-life of a parent chemical in soil can be considered differently from that in water or air. To control soil also means to control edible crops and animals in the area. If the chemical is immediately taken up and causes harm to plants or animals, then no half-life is acceptable. If a chemical is immediately taken up by plants and animals and metabolized without harmful breakdown products, and with no harm to plants and animals, then its presence may be acceptable. If the chemical has a half-life greater than 60 days, then the future use of the land must be considered for a chemical taken up by plants and animals. If the chemical breaks down in 1 hour, 1 day, 100 days, or longer and causes no environmental harm, then the half-life may not make a difference.

Air Compartment

This compartment is not usually studied; however, if a chemical is immediately harmful to plants and animals, no half-life is acceptable. Also, if it is not harmful but results in food-chain contamination, then no half-life is acceptable. Air cannot be controlled; so it is best to keep contaminating chemicals out of this compartment.

Plant and Animal Compartments

These compartments usually are studied for metabolism if the chemical is a pesticide. No discussions are presented herein on metabolism, as that subject is beyond the scope of this book.

The faster the hydrolytic rate is, the less likely the possibility of continued exposure in the environment. If the hydrolytic half-life ($t_{1/2}$ is <30 days, then accumulation, bioaccumulation, and foodchain contamination are not likely; if $t_{1/2}$ is between 30 and 90 days, chemical behavior goes either way; and if $t_{1/2}$ is > days, contamination is likely. Table 2-3 will help one to visualize this assessment.

Table 2-3. Hydrolytic Half-Life (Days) and Fate of Chemicals.

ENVIRONMENTAL COMPARTMENTS	RAPID $t_{1/2} < 30$	MEDIUM $t_{1/2}$ 30 TO 90	SLOW $t_{1/2} > 90$
Accumulation	not likely	either way	yes
Bioaccumulation	not likely	either way	yes
Food-chain contamination	not likely	either way	yes
Persistence	n *	either way	yes
Adsorption	n	either way	maybe
Dissipation	yes	either way	n to slowly
Problem if acutely hazardous	yes	either way	yes
Problem only if hazardous via long term	not likely	maybe	likely

* n denotes negligible.

REFERENCES

[1] Lyman, Warren J., William F. Reehl, and David H. Rosenblatt, *Handbook of Chemical Property Estimation Methods,* McGraw Hill Book Company, Chapter 7, 1982.

[2] Ney, Ronald E., Jr., "Fate, Transport and Prediction Model Application to Environmental Pollutants," Spring Research Symposium, James Madison University, UP, April 16, 1981.

PHOTOLYSIS

A chemical can be phototransformed as long as it can absorb sunlight. Phototransformation of a chemical can occur in air, soil, or water and on surfaces of water, soil, plants, and animals. The photolytic product(s) can be either a higher- or a lower-molecular-weight chemical(s). This process also has been called photodegradation. Photooxidation can occur in the stratosphere, but is not considered in this discussion.

A chemical has to sorb light to be phototransformed; thus, if a chemical can be analyzed spectrophotometrically, it has a good chance to be phototransformed. Environmental influences can have an effect on the rate of phototransformation, such as depth of the chemical in soil and in water, sorption to soil, sensitizers, quenchers, and pH. The rate of phototransformation is the time that it takes for a parent chemical to be transformed to one-half (the half-life, $t_{1/2}$) of its original amount. This rate could differ in soil, in water, and on surfaces. The rate of photolysis can be used to determine persistence in the environment.

The considerations used for hydrolysis also apply to photolysis. The faster the photolytic rate is, the less likelihood there is of continued exposure in the environment. If the photolytic half-life is < 30 days, then accumulation, bioaccumulation, or food-chain contamination is unlikely; if it is 30 to 90 days, the chemical behavior goes either way; and if it is >90 days, contamination is likely. Table 2-4 will help you to visualize the possibilities.

For a technical review of the photo process, Lyman et al. [1] go into explicit detail.

Table 2-4. Photolytic Half-Life (Days) and Fate of Chemicals.

ENVIRONMENTAL SURFACES	RAPID $t_{1/2} < 30$	MEDIUM $t_{1/2}$ 30 TO 90	SLOW $t_{1/2}$ >90
Accumulation	not likely	either way	yes
Bioaccumulation	not likely	either way	yes
Food-chain contamination	not likely	either way	yes
Persistence	n *	either way	yes
Adsorption	n	either way	maybe
Dissipation	yes	either way	slowly
Problem if acutely hazardous	yes	either way	yes
Problem only if hazardous via long term	not likely	maybe	likely

* n denotes negligible.

REFERENCE

[1] Lyman, Warren J., William F. Reehl, and David H. Rosenblatt, *Handbook of Chemical Property Estimation Methods,* McGraw Hill Book Company, Chapter 8, 1982.

VOLATILIZATION

A chemical on or in soil, on or in water, or on plants or animals may volatilize and get into the air. Vapor pressure is one of the most important factors governing volatilization, and provides an indication of whether a chemical will volatilize into the air under environmental conditions.

Some of the factors that effect volatilization in the environment are climate, sorption, hydrolysis, and phototransformation [1]. Here are guidelines:

1. A chemical with a low vapor pressure (VP), high adsorptive capacity, or high water solubility is less likely to volatilize into the air.
2. A chemical with a high VP, low sorptive capacity, or very low water solubility is more likely to volatilize into the air.
3. Chemicals that are gases at ambient temperatures will get into the air. (Gases at ambient temperature are not considered herein.)

Rapid volatilization into the air could result in immediate hazards for chemicals released indoors or outdoors if workers are in the area (e.g., agricultural workers). Vapor pressure is reported in terms of mm Hg (millimeter or mercury) or torr (which is equivalent to mm Hg). Table 2-5 will help you to visualize the possibilities.

REFERENCE

[1] Ney, Ronald E., Jr., "Fate, Transport and Prediction Model Application to Environmental Pollutants," Spring Research Symposium, James Madison University, UP, April 16, 1981.

SOIL SORPTION

The physical-chemical process by which a soil(s) ties up chemicals so that they are not released or are very slowly released in the environment is called adsorption, or bound residues. A chemical that is held by soil but is easily released is absorbed (in the way that a sponge holds water). The release mechanism considered herein is water, which is available in nature.

Plant root systems, as reported by Ney [4], have been shown to release adsorbed chemicals from soil for uptake into plant parts. Animals ingesting soil have also released the adsorbed chemical(s) for uptake. The mobility of adsorbed chemicals is prevented in soils. Here, the use of the word soil means only those soils that can adsorb chemicals, because soils do not all adsorb chemicals (e.g., sand, low-organic soils, etc., do not).

The movement of soil or soil particles containing an adsorbed chemical can contaminate other environments (e.g., as soil runoff or airborne particulates).

Sorption to soil, in almost all cases, can prevent phototransformation, hydrolysis, volatilization, mobility by water solubility, and microbial biodegradation.

Prior to the 1970s, many scientists would report that soils studied contained no chemical residues when, in fact, the scientists did not consider adsorbed residues or breakdown products. As previously discussed, the only way to discern what really occurs is to use a radiolabeled parent chemical. If the chemical cannot be extracted from soil with water

Table 2-5. Vapor Pressure (mm Hg or torr) and Fate of Chemicals.

ENVIRONMENTAL SURFACES	LOW < 0.000001	MEDIUM 0.000001 TO 0.01	HIGH > 0.01
Volatility	low	medium	high
Accumulation	yes	maybe	n *
Bioaccumulation	yes	maybe	n
Food-chain contamination	yes	maybe	n
Persistence	yes	maybe	n
Adsorption	high	maybe	low
Dissipation	yes	maybe	n
Problem if acutely hazardous	n	maybe	yes
Problem only if hazardous via long term	yes	maybe	n
Solubility	high	medium	low

* n denotes negligible.

or with other solvents, then there are two ways to discern the presence of a bound residue: (1) by combustion of the soil to measure adsorbed radioactivity, and (2) by planting crops to study any chemical residue(s) taken up by the crops (only needed if crops are involved).

Water is used as an extraction solvent to study bound residues. It should be noted, however, that other solvents, which do not occur naturally in the environment, may solubilize the adsorbed chemical for extraction. In most cases such solubilization will not happen, but it could be a problem in areas of solvent spills or in leaking storage areas.

Soil sorption, chemical sorption, or bound chemical(s) in soil may be expressed as the extent that an organic chemical partitions between a solid phase and a liquid phase. This value is better known as the adsorption coefficient (Koc), and is expressed as the μg adsorbed per organic carbon (soil solid phase) divided by μg per ml of solution (liquid base).

Authors have variously reported absorption coefficients or adsorption to organic matter as Kd, Koc, or Kom. Kom is the μg adsorbed per organic matter (soil organic) divided by μg per ml of solvent (liquid). This book will use only Koc.

Many relevant mathematical equations are given by Lyman et al. [2]. Helling and Turner [1] and Ney [3] have discussed soil sorption and leaching potential; Helling and Turner [1] have reported on soil thin layer chromatography (soil TLC). This is a very unique method, which could be used to combine many soil studies. Helling and Turner [1] divided soil mobility into classes, which will be discussed in a later chapter. These classes also can be used to predict adsorption. Comparing mathematical predictions, soil TLC, water solubility, and actual studies will help one to pinpoint erroneous scientific results.

We must remember that all soils do not act alike, nor do all chemicals. Thus, the chemical's identity and a study of soil characteristics are needed. Several soil characteristics may enhance sorption to a few chemicals, including the cation exchange capacity (CEC), clay-colloid-polymerization, clay with high surface areas, and so on.

To predict whether a chemical could be adsorbed to soil organic carbon (OC), one could use adsorption coefficients (Koc). The Koc value is only a numerical number with no units. Soils must contain organic carbon/organic matter to produce a Koc.

Chemical(s) with a high Koc of >10,000 will adsorb to OC. Chemicals with a Koc in the range of 1,000 to 10,000 could behave either way. Chemicals with a low Koc of 1,000 will not adsorb to soil OC. As discussed, these numbers for chemicals can be calculated or measured.

When they are coupled with soil TLC and other studies, a counter-check system is formed.

Soil TLC gives a measure of the movement through soil, expressed as Rf or TLC-Rf with no units even though measured as millimeters or lesser units.

Chemicals with a high TLC-Rf, > 0.75, will not be adsorbed in soil and will be mobile (leach). Chemicals with a TLC-Rf in the range of 0.34 to 0.75 could go either way. Chemicals with a low TLC-Rf, < 0.34, will be adsorbed and should not leach with water.

Table 2-6 helps us to visualize what can be predicted with a Koc and a TLC-Rf.

Chemicals that can be desorbed from soil are chemicals that were not adsorbed. The release of chemicals by water is called desorption. These chemicals usually have a WS in the range of >10 ppm and are mobile.

Table 2-6. Sorption to Soil and Fate of Chemicals.

SOILS AND SEDIMENTS	LOW KOC > 10,000 Rf< 0.34	MEDIUM KOC 1,000 TO 10,000 Rf 0.34 TO 0.75	HIGH KOC < 1,000 Rf > .75
Adsorption	yes	either way	n *
Mobility	n	either way	yes
Accumulation	yes	either way	n
Bioaccumulation	yes	either way	n
Food-chain contamination	yes	either way	n
Solubility	n	either way	yes
Persistence	yes	either way	n
Dissipation	n	either way	yes

* n denotes negligible.

REFERENCES

[1] Helling, Charles S. and Benjamin C. Turner, "Pesticide Mobility: Determination by Soil Thin-Layer Chromatography," *Science*, vol. 162, pp. 562-563, 1968.

[2] Lyman, Warren J., William F. Reehl, and David H. Rosenblatt, *Handbook of Chemical Property Estimation Methods*, McGraw-Hill Book Company, Chapter 4, 1982.

[3] Ney, Ronald E., Jr., "Fate, Transport and Prediction Model Application to Environmental Pollutants," Spring Research Symposium, James Madison University, UP, April 16, 1981.

[4] Ney, Ronald E., Jr., "Regulatory Aspects of Bound Residues," Workshop V-C, 4th International Congress of Pesticide Chemistry, International Union of Pure and Applied Chemistry, Zurich, Switzerland, UP, July 24-25, 1978.

LEACHING IN SOIL

The movement of a chemical downward through soil by water is called leaching. It is of concern because of the possibility that the chemical will move through the soil and contaminate the groundwater. If this happened, then well-water, aquatic-organism, and food-chain contamination could occur.

Many factors affect whether or not a chemical leaches in soil [1], including solubility, biodegradation, hydrolysis, dissociation, sorption, volatility, rainfall, and evapo-transpiration. Their effects on mobility are described as follows:

1. A chemical that is water-soluble can leach in soil and is likely to be biodegraded by soil microbes. If biodegradation is rapid, then leaching may not be a problem. Chemicals may be leached by solvents other than water, but other solvents are not considered herein as leaching solvents.
2. A chemical that is insoluble in water can be adsorbed in soil, moved with soil particles, and perhaps very slowly biodegraded, if at all.
3. Water-soluble chemicals can get into the air by evapotranspiration (as the water evaporates).
4. Chemicals that are highly volatile can get into the air and not be available for leaching.
5. Chemicals on the soil surface may be phototransformed and thus not be available to leach; however, the photoproducts may be available.
6. The more precipitation, the greater the chance is for chemicals to leach.
7. Hydrolysis and dissociation may prevent leaching.

The reader must remember that any parent compound that breaks down by any mechanism yields breakdown products that may or may not cause problems in the environment. The breakdown products should be regarded in the same light as the parent chemical, even though we are considering *only* parent compounds here for purposes of discussion and simplicity.

There are many ways to study or predict chemical movement through soil, as reported by Helling and Turner [l], Lyman et al. [2], and Ney [3, 4]. They include the use of:

1. Soil columns to measure chemical movement.
2. Solubility data to predict movement (see above).
3. Octanol water partition coefficient to indicate lipo-solubility, which is also an indicator of water solubility, in order to predict movement (see above).
4. Sorption coefficients to predict mobility (see above).
5. Soil TLC to predict mobility (see above) [l].

Table 2-7 shows what can be predicted with a *Koc,* a soil TLC-R*f,* and water solubility (WS).

Table 2-7. Leaching and Fate of Chemicals.

ENVIRONMENTAL COMPARTMENT: SOIL	$KOC > 10,000$ Rf < 0.34 ws < 10ppm	Koc 1,000-10,000 Rf 0.34-0.75 ws 10-1,000 ppm	Koc < 1,000 Rf > 0.75 ws > 1,000 ppm
Adsorbs	yes	either way	n *
Leaches	n	either way	yes
Biodegrades	n	either way	yes
Dissipates	n	either way	yes
Accumulates	yes	either way	n
Bioaccumulates	yes	either way	n
Causes food-chain contamination	yes	either way	n

* n denotes negligible.

REFERENCES

[1] Helling, Charles S. and Benjamin C. Turner, "Pesticide Mobility: Determination by Soil Thin-Layer Chromatography," *Science,* vol. 162, pp. 562-563, 1968.

[2] Lyman, Warren J., William F. Reehl, and David H. Rosenblatt, *Handbook of Chemical Property Estimation Methods,* McGraw-Hill Book Company, 1982.

[3] Ney, Ronald E., Jr., "Fate, Transport and Prediction Model Application to Environmental Pollutants," Spring Research Symposium, James Madison University, UP, April 16, 1981.

[4] Ney, Ronald E., Jr., "Exposure Assessment Considerations and Problems," Exposure Assessment Workshop, U.S. Environmental Protection Agency, Washington, D.C., UP, April 6-7, 1982.

RUNOFF

Runoff is the process by which a chemical is moved across a contaminated area to a noncontaminated area. The end result is contamination of the noncontaminated area. This occurs in a number of ways:

1. A water-soluble chemical can be solubilized in rainwater and snow melt, and as the water runs off the contaminated area, so does the soluble chemical.
2. Chemicals that are ad- or absorbed by soil can be moved with the soil as soil erosion occurs.
3. Solvents other than water may solubilize and move chemicals. (Remember that solvents other than water are also contaminates, but only water is considered as a solvent herein.)

When a chemical does run off or is moved with soil to a noncontaminated area, contamination of air, soil, plants, animals, and aquatic environmental compartments can result.

There are ways to predict whether runoff could occur. If chemicals are water-soluble, runoff is likely. Land surfaces are not always level; they may have slopes. Therefore, if the land has definite slopes that result in water runoff, it is likely that (1) soil is moved, and (2) any chemicals that are present are moved. One could even "eyeball" the movement of rainwater across land and predict runoff.

Using the WS, *Kow*, soil TLC-R*f*, or *Koc*, one can predict the movement of a water-soluble chemical across land. Table 2-8 shows how one might predict runoff.

Finally, it should be noted that chemical runoff may occur, to some extent, almost *100%* of the time if water runs off. The extent of the runoff may be small, as the newly contaminated area may be, but movement will occur. Of course, the amount of rainfall and snow strongly influences the amount and distance of such movement.

Table 2-8. Runoff and Chemical Movement.

SOIL SURFACE	WS < 10 ppm Rf < 0.34 KOC > 10,000 KOW > 1,000	WS 10-1,000 ppm Rf 0.34-0.75 KOC 1,000-10,000 KOW 500-1,000	WS > 1,000 ppm Rf > 0.75 KOC < 1,000 KOW < 500
Water runoff	n *	either way	yes
Soil runoff	yes	yes	yes

* n denotes negligible.

Chapter 3
Biological Processes

BIODEGRADATION

Biodegradation is the biological process by which aerobic microbes or anaerobic microbes break down organic chemicals to either a higher- or a lower-molecular-weight chemical(s) called a biodegradate(s). This is a very important process by which soil microbes or aquatic microbes can detoxify chemicals [2]. The process could also result in the formation of a more toxic chemical.

The biodegradation process can be evaluated by hydrolysis, phototransformation, leaching, and sorption studies. There are computer models that predict biodegradation, but at this time they have not been fully validated; and other estimation techniques are in the early stages, as reported by Lyman et al. [1].

Many factors may affect the rate of biodegradation. (Here the rate means the time required to break down the parent chemical through at least one-half of its original concentration.) Among the influences are pH, temperature, sorption, populations of microbes, different types of microbes, moisture, the presence of other chemicals, and the concentration of chemicals present.

Below are some guidelines that can be used to predict whether biodegradation can occur:

1. Chemicals that are highly water-soluble can biodegrade, but those with low WS usually will not.
2. Chemicals that adsorb in soil usually will not biodegrade, but those that do not adsorb can.
3. Chemicals with a high Kow usually will not biodegrade, but those with a low Kow can.

Table 3-1. Biodegradation of Chemicals

SOIL OR WATER	KOC >10,000 RF < 0.34 WS <10 PPM KOW > 1,000	KOC 1,000-10,000 RF 0.34-0.75 WS 10-1,000 PPM KOW 500-1,000	KOC <1,000 RF > 0.75 WS > 1,000 PPM KOW < 500
Adsorbs	yes	either way	n*
Leaches	n	either way	yes
Soluble	n	either way	yes
Dissipates	n to slowly	either way	yes
Accumulates	yes	either way	n
Bioaccumulates	yes	either way	n
Biodegrades	n to slowly	either way	yes

* n denotes negligible.

4. Chemicals that leach in soil usually will biodegrade, but those that do not leach usually will not.

Table 3-1 shows what can take place.

REFERENCES

[1] Lyman, Warren J., William F. Reehl, and David H. Rosenblatt, *Handbook of Chemical Property Estimation Methods,* McGraw-Hill Book Company, Chapter 9, 1982.
[2] Ney, Ronald E., Jr., "Fate, Transport and Prediction Model Application to Environment Pollutants," Spring Research Symposium, James Madison University, UP, April 16, 1981.

BIOACCUMULATION

If a chemical is taken up by a plant or an animal, the chemical can be metabolized—that is, changed to a higher- or lower-molecular-weight compound that may or may not be more toxic than the parent compound. There are other chemical degradation processes that could change a parent chemical (hydrolysis in animals, etc.), but these processes are not discussed as we only want to predict whether parent chemicals can build up, accumulate, or bioaccumulate in plants or animals.

Bioaccumulation is the key consideration, as we want to be able to predict whether food-chain contamination can occur. It may happen in

Table 3-2. Bioaccumulation of Chemicals

PLANTS AND ANIMALS	WS <10 PPM KOW > 1,000 RF < 0.34 KOC >10,000	WS 10-1,000 PPM KOW 500-1,000 RF 0.34-0.75 KOC 1,000-10,000	WS > 1,000 PPM KOW < 500 RF > 0.75 KOC < 1,000
Bioaccumulation	yes	either way	n *

* n denotes negligible.

two ways: (1) contaminated plants eaten by animals cause contamination in animals, and (2) contaminated animals eaten by animals contaminate animals. This is how contamination is passed up the food chain.

It is important to note that plant and animal metabolism studies are most complex, and studies on the metabolites formed may require new analytical methods and new toxicological studies.

As to the problem of predicting bioaccumulation, some of the predictive indicators are water solubility (WS) and the octanol water partition coefficient (Kow). Table 3-2 indicates how to predict whether bioaccumulation can occur.

Chapter 4
Exposure Assessment

There are many aspects to making exposure assessments. Exposure considerations should be based on the availability of a chemical residue to environmental compartments, with resultant exposure of wildlife and humans (grouped together as animals). Exposure to a chemical depends upon its environmental fate and transport and on how long it lasts in the environment.

The preceding discussions on water solubility, octanol water, hydrolysis, photolysis, volatilization, soil sorption, leaching, run-off, biodegradation, and bioaccumulation indicated whether chemicals would be available for exposure. It should be noted that there are other predictors, but I believe that they are beyond the scope of this discussion and should be left to qualified professionals. Lyman et al. [1], for example, present many other predictive techniques; and Ney [2] discussed considerations and problems associated with exposure assessment of hazardous waste constituents.

Table 4-1 indicates what could happen in the environment. Table 4-2 is a summary of predictive techniques and ranges, and of chemical fate and transport within these ranges, as given in previous chapters.

Exposure to some chemicals may not be harmful. If a chemical is acutely toxic, cancer-causing, mutagenistic, teratogenistic, and so on, then a qualified person should weigh its characteristics against the amount of the chemical available for exposure to animals. Equations and models have been devised for this but they are beyond the scope of this presentation. Remember that the purpose of this book is to help individuals to discern the fate and transport of chemicals in the environment, and to determine whether exposure is possible.

Table 4-1. Processes Affected by Environmental Characteristics.

ENVIRONMENT	PHYSICAL PROCESS	CHEMICAL PROCESS	METABOLIC PROCESS
Wind	Photolysis	Hydrolysis	Plant metabolism
Temperature	Sorption		Animal metabolism
Light	Desorption		Microbial metabolism
Soil	Volatility		
Runoff	Dissociation		
Leaching	Ion-exchange		

Table 4-2. Predicting Where the Chemical Goes.

PREDICTIVE TECHNIQUE	RANGES		
WS, ppm	< 10	10–1,000	> 1,000
Hydrolysis, $T_{1/2}$ days	> 90	30–90	< 30
Photolysis, $T_{1/2}$ days	> 90	30–90	< 30
VP,mm Hg	< 0.000001	0.000001–0.01	> 0.01
Koc	> 10,000	1,000–10,000	< 1,000
Rf	< 0.34	0.34–0.75	> 0.75
Kow	> 1,000	500–1,000	< 500
Soil, $T_{1/2}$ months	6	2–6	< 2
FATE AND TRANSPORT			
Soluble	n *	either way	yes
Hydrolyzes	n	either way	yes
Photolyzes	n *	either way	yes
Volatilizes	n	either way	yes
Adsorbs	yes	either way	n
Leaches	n	either way	yes
Runs off	n	either way	yes
Bioaccumulates	yes	either way	n
Persists	yes	either way	n
Biodegrades	slowly	either way	yes
Is metabolized	slowly	either way	yes

* n denotes negligible.

REFERENCES

[1] Lyman, Warren J., William F. Reehl, and David H. Rosenblatt, *Handbook of Chemical Property Estimation Methods,* McGraw-Hill Book Company, 1982.
[2] Ney, Ronald E., Jr., "Exposure Assessment Considerations and Problems," Exposure Assessment Workshop, U.S. Environmental Protection Agency, Washington, D.C., UP, April 6-7, 1982.

Chapter 5
Examples

This chapter presents some common chemicals along with some characteristic data, followed by a discussion of what may happen to each chemical in the environment, as well as exposure considerations.

There are several things to remember in using this material:

1. Refer to Table 4-2, on making predictions.
2. Remember that there always may be exceptions to the rules suggested by the table.
3. If two or more data points in Table 4-2 are satisfied, then a fit is likely for most points, in any of the three columns.
4. If two or more of a substance's physical, chemical, or biological properties fit a column in the table, then there is good likelihood that most of its other properties will fit the same column.
5. More data are available for most of the chemicals listed than are chosen for the example predictions. Remember that we may need to make predictions based on limited data.

There will be mention in the data, when available, of the bioconcentration factor, BCF, which is a chemical's ability to bioconcentrate or bioaccumulate in animals (fish, in this case). The data will be useful in discussing exposure. Refer to the appendix for an alphabetical listing of all chemicals included in this chapter.

CHEMICALS

Acenaphthene

Data

Pesticide

WS	3.47 mg/L	[4]
Kow	21,380	[4]
VP at 20°C	0.02 torr	[4]

Discussion

- WS indicates that this chemical could adsorb to soil, run off with soil, and be bioaccumulated, and it should not leach and should not be biodegraded.
- Kow indicates that this chemical should bioaccumulate and could cause food-chain contamination. Residues of this chemical could be expected in the food chain.
- VP indicates that this chemical could be volatile. If it is not phototransformed, fallout of the parent chemical could contaminate the food chain and water, and so could transformation products if they are formed.

Exposure

Routes of exposure could be by (1) ingestion of contaminated food and water and (2) inhalation of the volatile chemical.

Acenaphthylene

Data

WS	3.93 mg/L	[4]
Kow	11,749	[4]
VP	0.029 torr	[4]

Discussion

- WS indicates that this chemical could adsorb to soil, run off with soil, and be bioaccumulated, and it should not leach and should not be biodegraded.

- Kow indicates that bioaccumulation could occur and cause food-chain contamination. Residues of this chemical could be expected in the food chain.
- VP indicates that this chemical could be volatile. If it is not phototransformed, fallout of the parent chemical could contaminate the food chain and water, and so could phototransformation products if they are formed.

Exposure

Routes of exposure could be by (1) ingestion of contaminated food and water and (2) inhalation of volatile residues.

Acephate (O,S-Dimethyl acetylphosphoramidothioate)

Data

Pesticide		
WS	650,000 ppm	[6]
BCF—static water	0	[6]

Discussion

- WS indicates that this chemical should be biodegraded and could leach and run off, and that adsorption and bioaccumulation should not occur.
- BCF indicates that bioaccumulation and food-chain contamination should not occur, and is supported by the WS. BCF indicates that this chemical may be metabolized in animals.

Exposure

The routes of exposure could be by (1) inhalation if the chemical were volatile (but no data are presented on volatility), and (2) by ingestion of contaminated water and food if that occurred. Additional data would be needed for a more conclusive prediction.

Acetophenone

Data

Use—perfumery		[8]
Water solubility (WS)	5,500 ppm	[3]
Octanol water partition (Kow)	38.6 ± 1.2	[3]
Soil sorption coefficient (Koc)	35	[3]

Discussion

• WS is high; thus mobility should occur with water, and the chemical should be biodegraded by microbes.
• Kow is low, which indicates that bioaccumulation should not occur, which also is indicated by the high WS.
• Koc is low, which indicates that soil adsorption should not occur and that accumulation in soil should not occur.
• The three characteristics indicate that acetophenone should not be persistent in the environment. Bioaccumulation and foodchain contamination should not occur, based on the WS and the Koc. The WS, Koc, and Kow indicate that the chemical could leach and run off into water supplies if in soils that have no or limited microbial populations for biodegradation.

Exposure

Immediate exposure could be a problem if this chemical were acutely toxic. This chemical, if not degraded, could leach or run off into the aquatic environment and contaminate drinking water. Therefore, ingestion of contaminated drinking water could be a problem. If it were volatile, inhalation could be a problem, but no data are presented herein on volatility.

Acridine

Data

Use—manufacture of dyes		[3]
WS	38 ppm	[3]
Kow	4,200 ± 940	[3]
Koc	12,910	[3]
Volatile with steam		[3]

Discussion

- WS indicates that acridine should not be mobile (leached) and should not be biodegraded by microbes.
- Kow indicates that bioaccumulation and food-chain contamination are likely.
- Koc indicates that soil adsorption should occur, and this is supported by the low WS.
- Based on these aspects, the chemical acridine should not leach, could accumulate in soil, and could bioaccumulate and cause food-chain contamination. Runoff could occur with soil particles, resulting in aquatic contamination.

Exposure

Exposure could be through food-chain consumption. If the chemical causes cancer through long-term exposure, this could be a problem as long as the food is consumed. If it is in drinking water, the same exposure problem could exist. The chemical is volatile in steam and not at ambient temperatures; thus inhalation of this chemical should not be a problem at ambient temperatures.

Acrylonitrile (also Vinyl cyanide)

Data

WS	73,500 mg/L	[4]
VP at 22.8°C	100 torr	[4]
Kow	1.38	[4]
Degraded by microbes		[4] .

Discussion

- WS indicates that runoff, leaching, and biodegradation (as reported) could occur. The chemical should not bioaccumulate or adsorb in soil.
- VP indicates volatility; thus contamination to water, crops, and animals could occur via fallout. Inhalation could be a problem.
- Kow indicates that bioaccumulation should not occur.
- Biodegradation by microbes could occur.

Exposure

Routes of exposure could be by (1) inhalation of volatile residue; (2) ingestion of contaminated water, plants, and animals if the chemical precipitated out of the atmosphere; and (3) ingestion of water if the chemical leached to the drinking water supply.

Alachlor (2-Chloro-2′,6′ -diethyl-N-(methoxymethyl) acetanilide) (also Lasso)

Data

Pesticide		
WS	242 ppm	[6]
Koc	190	[6]
Kow	830	[6]
BCF—static water	0	[6]

Discussion

- WS indicates that this chemical could be either mobile or not, be bioaccumulated or not, and be adsorbed or not. It is predicted that the chemical should be mobile.
- Koc indicates some adsorption.
- Kow indicates that bioaccumulation is possible.
- BCF indicates that bioaccumulation should not occur, and that metabolism in animals could occur, based on the zero BCF number.
- The chemical could be mobile and get into water. Biodegradation may occur in soils and prevent leaching; however, no data are presented to support this prediction.

Exposure

Exposure could be by routes of (1) drinking contaminated water if mobility occurred, (2) inhalation if the chemical were volatile, and (3) ingestion of contaminated food if that occurred. No data are presented to support these predictions.

Aldicarb (2-methyl-2-(methylthio) propionaldehyde O-(methylcarbanoyl) oxime)

Data

Pesticide
WS	7,800 ppm	[6] [9]
Kow	11.02	[9]
Koc	0.073	[9]
BCF—static water	42	[6]

Discussion

- WS indicates that this chemical could leach into groundwater (it has done so) and could run off. It could be biodegraded in soils with good microbial populations and high organic matter contents.
- Kow indicates that this chemical should not bioaccumulate, and this is supported by the high WS.
- Koc indicates this chemical should not adsorb to soil, and this is supported by the high WS and low Kow.
- BCF is indicative that residues may be found in fish, but bio-accumulation should not occur based on the BCF number and the Kow number. This chemical is acutely toxic; thus fish would probably die. Ingestion of contaminated drinking water could be a problem.

Exposure

Immediate exposure is a problem, as this chemical is acutely toxic. If drinking water is contaminated, one must consider what is a safe level for drinking and what levels may be a danger to aquatic organisms and animals that eat the organisms (dead or alive). If this chemical reaches soils with low organic matter content and low microbial populations, leaching could occur and has occurred. Inhalation could be a problem if this chemical were found to be volatile, but no volatility data are presented herein.

Aldrin (Hexachlorohexahydro-endo, exo-dimethanonaphthalene 95% and related compounds 5%)

Data

Pesticide		
WS	0.013 ppm	[6]
Koc	410	[6]
BCF—static water	3,140	[6]

Discussion

- WS indicates that leaching and biodegradation should not occur, and that adsorption and bioaccumulation could occur. Runoff with soil particles could occur.
- Koc indicates that adsorption should not occur to any great extent, but it does occur, although perhaps not enough to prevent leaching to groundwater. The Koc value is contradictory to WS and BCF values, which indicate adsorption to soil.
- BCF indicates bioaccumulation and food-chain contamination. Residues could be expected in the food chain.
- WS and BCF indicate soil adsorption and bioaccumulation, as well as the potential for residues to be in the food chain. Koc indicates just the opposite, and indicates potential for leaching. Additional data would be needed to give a more precise prediction. However, two values do support each other and can be used until proven otherwise (i.e., the WS and BCF values).

Exposure

Routes of exposure could be by (1) ingestion of contaminated food and water and (2) ingestion of contaminated food. If the chemical were volatile, inhalation could be a problem; however, no data are presented herein on volatility. Food-chain contamination could be a major problem.

Amitrole (3-Amino-s-triazole) (also 3-AT)

Data

Pesticide		
Soil TLC-Rf	0.73	[1]

Discussion

- Soil TLC indicates leaching; thus this chemical should be water-soluble, biodegraded, and mobile, and it should not be adsorbed or bioaccumulated.
- With only one piece of data any prediction could be entirely wrong; however, when a prediction is required, one must be given with explanations and disclaimers as needed. In this case, additional data would be needed to support any predictions.

Exposure

Routes of exposure could be by (1) drinking contaminated water and (2) ingestion of contaminated food. If the chemical were volatile, inhalation could be a problem; however, no data are presented herein on volatility. Additional data would be needed to support and verify any of these predictions.

Ammonia

Data

Used in refrigeration and many other uses	[8]
Gas	[8]

Discussion

- If ammonia gas were released into the environment, inhalation could be a problem. Such releases could occur from refrigeration units that use ammonia, manufactures, industries, and water purification processing. Phototransformation may be a problem; however, there is no indication of such.

Exposure

Route of exposure could be by inhalation of volatile residues.

Aniline

Data

Pesticide	
Used in manufacturing of dyes, perfumes, and medicinals	[8]

WS	36,600 ppm	[6]
Kow	7	[6]
BCF—static water	6	[6]

Discussion

- WS indicates that this chemical could leach, run off, and be biode-graded, and it should not adsorb to soil and should not be bioaccumulated.
- Kow indicates that this chemical should not bioaccumulate.
- BCF indicates that bioaccumulation should not be a problem; how-ever, there is a possibility of residues in the food chain. The possibil-ity of metabolism in animals also is indicated, based on the low BCF value.

Exposure

Routes of exposure could be by (1) ingestion of contaminated food and water if such occurred, and (2) inhalation if the chemical were volatile, but no data on volatility are presented herein.

Anthracene (also Paranaphthalene)

Data

Pesticide

WS	0.073 ppm	[6]
Koc	26,000	[6]
Kow	22,000	[6]

Discussion

- WS indicates that leaching and biodegradation should not occur, and that adsorption, bioaccumulation, and runoff with soil particles could occur.
- Koc indicates adsorption to soil, and this should prevent leaching.
- Kow indicates that bioaccumulation and food-chain contamination could occur. This prediction is supported by WS and Koc values.

Exposure

Exposure routes could be by (1) ingestion of contaminated food and water and (2) inhalation if volatility occurred; however, no data are presented on volatility. If this chemical got into the aquatic environment, it could bioaccumulate and cause food-chain contamination. There are indications that consumption of contaminated water and food could result in chemical residues in animals. If it is also possible that bioaccumulation in plants could occur, and consumption of contaminated plants could also cause food-chain contamination.

Asbestos

Data

Naturally occurring and has many uses
WS—not soluble	[4]
Suspended in water	[4]
Not biodegraded	[4]
Not chemically degraded	[4]
Not volatile	[4]
Picked up by air as an aerosol	[4]
Not photolyzed	[4]
Can absorb other chemical	[4]

Discussion

- Asbestos is stable in the environment; therefore, it is resistant to chemical breakdown, biodegradation, phototransformation, and probably metabolism. Dissipation is unlikely in the environment.
- Asbestos fibers can be physically broken down to smaller fibers.
- Asbestos can be airborne; therefore, inhalation of the fibers is a problem. Airborne asbestos fibers can fall out to contaminate water, and thus be suspended in the water; they should not volatilize or break down in water, but should be persistent in water, and could be a problem in fish gills. Fallout of airborne asbestos fibers can also contaminate surfaces, such as food.

Exposure

Routes of exposure could be by (1) consumption of contaminated water with suspended asbestos particles, (2) consumption of food with asbestos on its surface, and (3) inhalation of asbestos airborne fibers.

Atrazine (2-Chloro-4-ethylamino-6-isopropylamino-s-triazine)

Data

Pesticide

WS	33 ppm	[2] [6]
VP	1.4 mm Hg × 10^6 at 30°C	[2]
Koc	149	[6]
Kow	476	[6]
BCF—flowing water	0	[6]
BCF—static water	11	[6]
Biodegraded by microbes		[2]
Soil TLC-Rf	0.47; class 3	[5]

Discussion

- WS indicates some solubility; thus biodegradation may occur, and this has been reported.
- VP indicates volatility.
- Koc indicates that adsorption should not occur in soil.
- Kow indicates that bioaccumulation should not occur.
- Soil TLC-Rf of 0.47 in class 3 does indicate that mobility is possible; thus the WS of 33 ppm is supported by mobility class 3. This chemical may get into water.
- Based on these data, volatility may be of more concern than mobility through or across soils.

Exposure

Exposure could occur if this chemical reached groundwater, or surface water, resulting in the contamination of drinking water. Biodegradation may be too slow to prevent leaching, and adsorption to soil should not occur. If this chemical finds its way to sandy soils, low-organic soils, and soils with low microbial content, then leaching is likely. In any case, the toxicity of this chemical may be of concern, and that should be looked into if drinking water is contaminated. Because of its volatility, inhalation exposure should be of concern. Fallout of volatile residues into noncontaminated areas could be a problem unless Atrazine is phototransformed; thus food-chain contamination is possible.

Benefin (also *N*-Butyl-*N*-ethyl-alpha, alpha, alpha-trifluoro-2, 6-dinitro-*p*-toluidine)

Data

Pesticide

WS	< 1 ppm	[6]
Koc	10,700	[6]

Discussion

- WS indicates that benefin should not leach in soil, but could run off with soil particles.
- Koc indicates that benefin should accumulate, and adsorb in soils.
- Based on the WS and the Koc, benefin may bioaccumulate in the food chain and not be or only slowly be biodegraded.
- If this chemical gets into water, it may be persistent, and it could bioaccumulate, resulting in food-chain contamination.

Exposure

Exposure could be by crop uptake, if crops are grown in contaminated soils, with the chemical passed on to animals that eat the food. Were the chemical to get into water, food-chain contamination could occur. From these two pieces of data, exposure would be by ingestion, and this would not preclude other exposure routes if more data were given, such as inhalation.

Bentazon (3-Isoprophyl-1*H*-2,1,3-benzothiadizen-4 (3*H*)-one 2,2-dioxide) (also Basagran)

Data

Pesticide

WS	500 ppm	[6]

Koc	0	[6]
Kow	220	[6]
BCF—static water	0	[6]

Discussion

- WS indicates that the chemical could run off, leach, and be biodegraded, and it should not bioaccumulate or adsorb to soil.
- Koc indicates that adsorption should not occur.
- Kow indicates that the chemical should not bioaccumulate, and food-chain contamination should not occur.
- BCF indicates that bioaccumulation should not occur, and that residues should not be expected in animals. The BCF value of zero indicates that the chemical is metabolized in animals. Koc and Kow values support these predictions.

Exposure

Routes of exposure could be by (1) ingestion of contaminated water, if such occurred; (2) inhalation, if the chemical were volatile, but no data on volatility are presented herein; and (3) ingestion of contaminated food, if such occurred.

Benzene

Data

WS	1,780 ppm	[6]
Koc	83	[6]
Kow	135	[6]
VP	95.2 torr	[4]
VP	76 mm Hg	[7]

Discussion

- WS indicates that benzene could be mobile in soils, and may be biodegraded.
- Koc indicates that adsorption should not occur.
- Kow indicates that bioaccumulation should not occur.
- VP indicates that benzene could volatilize and get into air. Volatilization should help to prevent leaching in soils.

- Benzene could leach, volatilize, and be biodegraded by microbes. Bioaccumulation should not be a problem; thus food-chain contamination is unlikely. Inhalation could be a problem.

Exposure

Exposure routes can be predicted to be (1) inhalation due to volatilization and (2) drinking contaminated water if benzene gets to drinking water. Fallout of volatile residues could be a problem if benzene is not phototransformed.

Benzidine

Data

WS	400 mg/L	[4]
Adsorbs to clay particles		[4]
Absorbs light	287 to 340 nm	[4]
Kow	64	[4]

Discussion

- WS indicates that the chemical could run off, leach, and be biodegraded.
- This chemical adsorbs to clay particles and not to organic matter (which is an unusual process, but still adsorption). This could not be predicted based on the data used herein.
- The chemical absorbs light, which indicates that the chemical could be phototransformed to other products.
- Kow indicates that bioaccumulation is unlikely.

Exposure

Routes of exposure could be by drinking contaminated water if leaching and runoff occurred. It cannot be predicted whether adsorption to clay is significant enough to prevent leaching because no data are presented. Inhalation could be a problem, if the chemical were volatile, as could phototransformation products; however, no data are presented to make such predictions.

Benzo [a] anthracene

Data

WS at 25°C	0.014 mg/L	[4]
Kow	407,380	[4]
VP at 20°C	5×10^{-9} torr	[4]

Discussion

- WS indicates that this chemical could adsorb to soil, run off with soil, and be bioaccumulated, and it should not leach and should not be biodegraded.
- Kow indicates bioaccumulation and potential for food-chain contamination. Residues could be expected in the food chain.
- VP indicates that this chemical should not be volatile.

Exposure

Routes of exposure could be by ingestion of contaminated food and water if this chemical were released in the environment.

Benzo [b] fluoranthene

Data

WS at 25°C	0.012 mg/L	[4]
Kow	3,715,352	[4]
VP at 20°C	5×10^{-7} torr	[4]

Discussion

- WS indicates that this chemical could absorb to soil, run off with soil, and be bioaccumulated, and it should not leach and should not be biodegraded.
- Kow indicates bioaccumulation and potential for food-chain contamination. Residues could be expected in food and water.
- VP indicates that this chemical should not be volatile.

Exposure

Routes of exposure could be by ingestion of contaminated food and water if this chemical were released into the environment.

Benzo [a] pyrene

Data

WS at 25°	0.0038 mg/L	[4]
Kow	1,096,478	[4]
VP at 20°	5×10^{-9} torr	[4]

Discussion

- WS indicates that the chemical could adsorb in soil, run off with soil, and be bioaccumulated, and it should not leach and should not be biodegraded.
- Kow indicates bioaccumulation and potential food-chain contamination. Residues could be expected in food and water.
- VP indicates that volatility is of no concern.

Exposure

Routes of exposure could be by ingestion of contaminated food and water.

Bifenox (Methyl 5-(2,4-dichlorophenoxy)-2-nitrobenzoate) (also Modown)

Data

WS	0.35 ppm	[6]
BCF—static water	200	[6]

Discussion

- WS indicates that this chemical should adsorb to soil, run off with soil, and be bioaccumulated, and it should not leach and should not be biodegraded.
- BCF indicates that this chemical could bioaccumulate and could cause food-chain contamination. Residues could be expected in the food chain if the chemical were released in the environment.

Exposure

Routes of exposure could be by (1) consumption of contaminated water and food if such occurred, and (2) inhalation if the chemical were volatile; however, no data are presented herein on volatility.

Biphenyl (also Diphenyl)

Data

Pesticide		
WS	7.5 ppm	[6]
Kow	7,540	[6]
BCF—flowing water	340	[6]

Discussion

- WS indicates that this chemical should not leach and not be biodegraded; however, it could adsorb to soil, run off with soil, and be bioaccumulated.
- Kow indicates that bioaccumulation should occur.
- BCF indicates that a residue could be present, and that this chemical should not bioaccumulate. This contradicts what the WS and Kow values indicate. BCF indicates that residues could be found in animals; however, metabolism should reduce the amount of residue found.
- WS and Kow are in contradiction with BCF, and this indicates that additional testing is needed; however, BCF does indicate that there may be a problem with residues in the food chain, though not bioaccumulated. Metabolism of the chemical in animals could occur, but it might be at a very slow rate.

Exposure

Routes of exposure could be by (1) ingestion of crops, animals, and water if they were contaminated, and (2) inhalation of volatile residue if such occurred; however, no data are presented on volatility.

Bromacil (5-Bromo-3-*sec*-butyl-6-methyluracil)

Data

Pesticide
WS	8.5 ppm	[6]
Koc	72	[6]
Soil TLC-Rf	0.69	[1]

Discussion

- WS indicates that the chemical should be adsorbed to soil, run off with soil particles, and be bioaccumulated, and it should not be biodegraded and should not be leached.
- Koc indicates that adsorption to soil should not occur; however, it could occur in small amounts.
- Soil TLC-Rf indicates that the chemical could leach.
- The three pieces of data are contradictory; however, Koc and soil TLC indicate that leaching could occur, and that is what should be predicted.

Exposure

Routes of exposure could be by (1) drinking contaminated water, (2) ingestion of contaminated food, and (3) inhalation if volatility occurred; however, no data are presented herein to indicate volatility. Because of the contradiction betweem WS, Koc, and soil TLC, additional data would be needed for a more conclusive prediction of what could happen to this chemical in the environment.

Bromobenzene

Data

WS	446 ppm	[6]
Kow	900	[6]

Discussion

- WS indicates that this chemical could go either way in relation to leaching, runoff, adsorption, degradation, and bioaccumulation.

- Kow indicates that this chemical could bioaccumulate or be slowly metabolized. Residues could be expected in the food chain.
- There are not enough data for one to make a precise prediction with the data given. However, a good prediction, using the limited data given, would be that the chemical could result in the contamination of water and the food chain, and residues could be expected in the food chain.

Exposure

Routes of exposure could be by (1) consumption of contaminated water and food if such occurred, and (2) inhalation if the chemical were volatile; however, no data are presented herein on volatility.

Bromodichloromethane

Data

Kow	76	[4]
VP at 20°C	50 torr	[4]

Discussion

- Kow indicates that this chemical should not bioaccumulate; thus it is also predicted that it could leach, run off, and be biodegraded, and it should not be adsorbed.
- VP indicates that the chemical should be volatile; thus inhalation could be a problem. If it were phototransformed, photoproducts and the parent chemical could fall out and contaminate water and food.
- This chemical could be mobile in the environment, and could contaminate water and food, and thus could result in contamination of the food chain. Volatile residues could present an inhalation problem.

Exposure

Routes of exposure could be by (1) consumption of contaminated water and food if such occurred, and (2) inhalation of volatile residues.

Bromoform (also Tribromomethane)

Data

WS at 30°C	3,190 mg/L	[4]
Kow	200	[4]
VP at 34°C	10 torr	[4]

Discussion

- WS indicates that this chemical should leach, run off, and be biodegraded, and it should not adsorb and should not be bioaccumulated.
- Kow indicates that the chemical should not bioaccumulate, and this is supported by the value given for WS.
- VP indicates that volatility should occur for this chemical. If it were phototransformed, both the parent chemical and photoproducts could fall out and contaminate water and food. Inhalation could be a problem.
- This chemical is mobile and could contaminate water and food, thus resulting in contamination of the food chain. Residues in the environnment could dissipate by biodegradation and metabolism; however, no data are presented on these mechanisms. Inhalation of volatile residues could be a problem.

Exposure

Routes and exposure (1) could be by consumption of contaminated water and food, if such occurred, and (2) should be by inhalation of volatile residues.

Bromomethane (also Methyl bromide)

Data

WS	900 mg/L	[4]
Kow	12.59	[4]
VP at 20°C	1,420 torr	[4]

Discussion

- WS indicates that this chemical could go either way, but Kow indicates that it should not bioaccumulate; therefore, it is predicted that it should leach, run off and biodegrade, and it should not adsorb to soil and should not be bioaccumulated.
- Kow indicates that bioaccumulation should not occur. It is possible that residues of this chemical could be found in water and food; however, its volatility should help prevent this.
- VP indicates that this chemical is volatile, and this could cause inhalation problems. If it were phototransformed, fallout of photoproducts and the parent chemical could contaminate the environment.
- This chemical is volatile and could present inhalation problems with the parent chemical and potential photoproducts. Although residues could be present in the environment, it is predicted that they would not be there long because they would volatilize.

Exposure

Route of exposure should be by inhalation of volatile residues.

4-Bromophenyl phenyl ether

Data

WS at 20°C	38 mg/L	[4]
Kow	141,254	[4]
VP at 20°C	0.0015 torr	[4]

Discussion

- WS indicates that this chemical could adsorb to soil, run off with soil, and be bioaccumulated, and that it should not leach and should not be biodegraded.
- Kow indicates bioaccumulation and potential for food-chain contamination. Residues could be expected in the food chain.
- VP indicates potential for volatility. If volatility occurred, fallout could cause contamination of water and the food chain. If the chemical were phototransformed, transformation products also could contaminate the food chain and water.

Exposure

Routes of exposure could be by (1) ingestion of contaminated food and water and (2) inhalation of volatile residues.

Bufencarb (3- (1-Methylbutyl) phenyl methylcarbamate) (also Bux, Metalkamate, etc.

Data

Pesticide
WS   < 50 ppm   [6]
BCF—static water   0   [6]

Discussion

- WS indicates that this chemical could go either way in relation to leaching, runoff, adsorption, biodegradation, and bioaccumulation.
- BCF indicates that the chemical should not bioaccumulate.
- This chemical should not bioaccumulate, but could be mobile enough to contaminate the food chain. If this occurred, it should be metabolized and biodegraded. There is a possibility that residues could exist in the food chain if dissipation were not rapid enough.

Exposure

Routes of exposure could be by (1) consumption of contaminated water and food if such occurred, and (2) inhalation if the chemical were volatile; however, no data are presented herein on volatility.

Butralin (4-(1,1-Dimethyl) -N-(1-methylpropyl) -2, 6-dinitrobenzenamine) (also Sector)

Data

WS   1 ppm   [6]
Koc   8,200   [6]

Discussion

- WS indicates that this chemical should be adsorbed to soil, and be bioaccumulated and runoff with soil, and it should not leach and should not be biodegraded.
- Koc indicates that this chemical should be adsorbed to soil and could be bioaccumulated; this is supported by the water solubility value.

Exposure

Routes of exposure could be by (1) consumption of contaminated water and food if such occurred, and (2) inhalation if the chemical were volatile; however, no data are presented herein on volatility.

Butyl benzyl phthalate

Data

WS	insoluble	[4]
Kow	26,301	[4]

Discussion

- WS indicates that this chemical could bioaccumulate, adsorb to soil and run off with soil particles, and it should not be biodegraded and should not be leached.
- Kow indicates that it could bioaccumulate and cause food-chain contamination.

Exposure

Routes of exposure could be by (1) ingestion of contaminated food or water and (2) inhalation if the chemical were volatile; however, there are no data presented to indicate volatility.

x-sec-Butyl-4-chlorodiphenyloxide

Data

WS	0.14 ppm	[6]
Kow	16,000	[6]
BCF—flowing water	298	[6]
BCF—static water	400	[6]

Discussion

- WS indicates that this chemical should adsorb in soil, run off with soil, and be bioaccumulated, and it should not be leached and should not be biodegraded.
- Kow indicates that the chemical should be bioaccumulated and could cause contamination to the food chain.
- Both BCF values indicate that the chemical should be in the food chain; however, bioaccumulation may be prevented by metabolism in animals. No matter which occurs, high residues should be expected in animals.
- This chemical should be expected to be present in the food chain if released into the environment. Residues also should be expected in the food chain; however, they may be slowly metabolized in animals.

Exposure

Routes of exposure (1) should be by consumption of contaminated water and food if such occurred, and (2) could be by inhalation if the chemical were volatile; however, no data are presented herein on volatility.

Captan (N-trichloromethylthio-4-cyclohexene-1,2-dicarboximide) (also Orthocide and Vancide 89)

Data

Pesticide
WS	< 0.5 ppm	[6]
Kow	224	[6]
BCF—static water	0	[6]
Soil TLC-Rf	0.39	[1]

Discussion

- WS indicates that this chemical should be adsorbed in soil, be bioaccumulated, and run off with soil, and it should not leach and should not be biodegraded.
- Kow indicates that bioaccumulation should not occur.
- BCF indicates that bioaccumulation should not occur, and that metabolism in animals should occur.

- Soil TLC-Rf indicates that there is a possibility of leaching. There is a contradiction between WS, and Kow, BCF, and soil TLC. However, one could predict that leaching is possible, and that residue could be found in the food chain.

Exposure

Routes of exposure could be by (1) drinking contaminated water, (2) ingestion of contaminated food, and (3) inhalation if the chemical were volatile; however, no data are presented on volatility herein.

Carbaryl (1-Naphthyl methylcarbamate) (also Sevin)

Data

Pesticide

WS	40 ppm	[6]
Koc	230	[6]
Kow	230	[6]
Soil TLC-Rf	0.38	[1]
BCF	< 1	[6]

Discussion

- WS indicates that leaching and runoff could occur.
- Koc shows that adsorption should not occur.
- Kow shows that accumulation should not occur.
- Soil TLC-Rf indicates that leaching could occur but should be slight.
- BCF indicates that food-chain bioaccumulation should not occur.
- Carbaryl should biodegrade based on solubility, could leach and run off, but should not accumulate in soil or bioaccumulate; thus food-chain contamination is unlikely.

Exposure

Exposure would be by direct contact or by consumption of contaminated food or contaminated drinking water if carbaryl reached drinking water prior to biodegradation.

Carbofuran (2,3-Dihydro-2,2-dimethyl-7-benzofuranyl methylcarbamate) (also Furadan)

Data

Pesticide		
WS	415 ppm	[6]
Kow	40	[6]
BCF—static water	0	[6]

Discussion

- WS indicates that the chemical could leach, run off, and be biode-graded, and it should not be adsorbed and should not be bioaccumulated.
- Kow indicates that bioaccumulation is unlikely.
- BCF indicates that bioaccumulation is unlikely.
- This chemical has caused bird kills; thus it may cause death in some animals prior to any bioaccumulation.

Exposure

Routes of exposure could be by (1) drinking contaminated water, (2) ingestion of contaminated food, and (3) inhalation if the chemical were volatile; however, no data on volatility are presented herein. This chemical has been known to cause bird kills; thus residues in the food chain are possible.

Carbon tetrachloride (or Tetrachloroethane)

Data

WS	785 ppm	[4]
VP at 20°C	90 torr	[4]
Hydrolysis	$t_{1/2}$ 7,000 years	[4]
Absorbs short wavelengths of light and higher-energy UV in the stratosphere		[4]

Discussion

- WS indicates that leaching, runoff, and biodegradation could occur.

- VP indicates that volatility could occur.
- Hydrolysis indicates that this chemical should not hydrolyze in water to be of any degradation benefit. It should be stable in environments without microbes.
- Phototransformation is unlikely to occur in the breathing atmosphere; however, it could occur in the stratosphere.
- Carbon tetrachloride could be available for inhalation and be mobile enough to get into water if biodegradation is slow. Bioaccumulation is unlikely, as is food-chain contamination.

Exposure

Exposure could be by two routes: (1) inhalation due to the chemical's volatility and (2) drinking contaminated drinking water. Acute dangers due to inhalation should be evaluated. Fallout of volatile residues to noncontaminated areas could be a problem, especially to aquatic environments, as this chemical should not be phototransformed.

Chloramben (3-amino-2,5-dichlorobenzoic acid) (also Amiben)

Data

Pesticide

WS	700 ppm	[6]
Koc	21	[6]

Discussion

- WS indicates that this chemical could go either way in relation to leaching, adsorption, runoff, biodegradation, and bioaccumulation.
- Koc indicates that adsorption is negligible; therefore, it is predicted that this chemical could leach, runoff and be biodegraded, and should not be absorbed and not be bioaccumulated.

Exposure

Routes of exposure could be by (1) consumption of contaminated water and food if such occurred, and (2) inhalation if the chemical were volatile; however, no data are presented herein on volatility. Additional data are needed for more precise prediction.

Chlorbromuron (3-(4-bromo-3-chlorophenyl)-1-methoxy-1-methylurea)

Data

Pesticide
WS	50 ppm	[6]
Koc	460	[6]
Soil TLC-Rf	0.14	[1]

Discussion

- WS indicates that this chemical could leach, run off, and be biodegraded, and it should not adsorb to soil and should not be bioaccumulated.
- Koc indicates that adsorption to soil should not occur; however, some adsorption is possible.
- Soil TLC-Rf indicates that leaching is unlikely, which contradicts the WS; however, perhaps biodegradation had occurred in this case, or maybe adsorption had taken place. Remember that the WS is in the "either way" range.
- Bioaccumulation of residues in the food chain may be possible, based on Koc and soil TLC values.

Exposure

Routes of exposure could be by (1) drinking contaminated water, (2) ingestion of contaminated food, and (3) inhalation if the chemical were volatile; however, no data are presented herein to indicate volatility.

Chlordane, Technical (60% Octachloro-4,7-methanotetrahydroindane and 40% related compounds)

Data

Pesticide—termite control
WS	0.056 ppm	[6]
BCF—flowing water	11,400	[6]
BCF—static water	3,140	[6]
Hydrolysis—loses Cl in presence of alkaline reagents		[8]

Discussion

- WS indicates that chlordane should not be biodegraded in soils, should adsorb to soil, should accumulate in soil, and should be persistent in the soil environment.
- BCF indicates that it should bioaccumulate and cause food-chain contamination.
- Hydrolysis is likely in alkaline soils.
- Chlordane is persistent and can bioaccumulate in the environment and be taken up by plants and animals. It could persist in water, and if it gets into the aquatic environment, food-chain contamination is likely.

Exposure

Exposure could be by (1) drinking contaminated water or (2) eating contaminated plants and animals. If this chemical were volatile, than fallout to noncontaminated areas would be of concern, as well as phototransformation. Inhalation of volatile residues could then be a problem. Bioaccumulation and food-chain contamination are likely if this chemical is released to the environment.

Chlorine

Data

Used in water purification and many other uses   [8]
Gas   [8]

Discussion

- If released, chlorine could cause an inhalation problem. It also can combine with other chemicals in the air to form acids, which may present environmental problems.

Exposure

Routes of exposure could be by (1) inhalation of volatile residues and (2) inhalation of acids that might be formed in air.

Chlorobenzene

Data

WS at 20°C	almost insoluble	[4]
Kow	692	[4]
VP at 20°C	8.8 torr	[4]

Discussion

- WS indicates that this chemical could adsorb to soil, run off with soil, and be bioaccumulated, and it should not leach and should not be biodegraded.
- Kow indicates that perhaps bioaccumulation could occur, and residues could be expected in the food chain.
- VP indicates volatility, and fallout of volatile chemicals could contaminate food and water. If the chemical were phototransformed, transformation products could contaminate food and water. Inhalation of the parent compound and its potential photoproducts could be a problem.

Exposure

Routes of exposure could be by (1) ingestion of contaminated food and water if such occurred, and (2) inhalation of volatile residues.

4-Chlorobiphenyl

Data

WS	1.65 ppm	[6]
Kow	79,400	[6]
BCF—flowing water	590	[6]

Discussion

- WS indicates that this chemical could be adsorbed to soil, run off with soil, and be bioaccumulated, and it should not leach and should not be biodegraded.
- Kow indicates that this chemical should bioaccumulate and cause food-chain contamination.

- BCF indicates that this chemical should bioaccumulate. Residues could be expected in the food chain if contamination occurred. The chemical appears to have potential for metabolism in animals; however, this should be very slow, based on the BCF value.
- If this chemical were to get into water, food-chain contamination could be expected.

Exposure

Routes of exposure could be by (1) consumption of contaminated water and food if such occurred, and (2) volatility if the chemical were volatile; however, no data are presented herein on volatility.

p-Chloro-m-cresol

Data

Antiseptic and disinfectant		[8]
WS at 20°C	3,850 mg/L	[4]
Kow	891	[4]
Aqueous solution turns yellow when exposed to light and air		[8]
Volatile in steam		[8]

Discussion

- WS indicates that this chemical could leach, run off, and be biodegraded, and it should not adsorb in soil and should not be bioaccumulated.
- Kow indicates that maybe bioaccumulation could occur, which contradicts WS. In either event, residues could be expected in the food chain.
- Exposure of this chemical in aqueous solution to light and air could transform it to other products.
- Because the chemical is volatile in steam, volatility would not be expected at ambient temperatures.

Exposure

Routes of exposure could be by (1) consumption of contaminated drinking water if this chemical leached or ran off into water, and (2) ingestion of contaminated food.

4-Chlorodiphenyloxide

Data

WS	3 ppm	[6]
Kow	12,000	[6]
BCF—flowing water	736	[6]

Discussion

- WS indicates that this chemical could be adsorbed to soil, run off with soil, and be bioaccumulated, and it should not be leached and should not be biodegraded.
- Kow indicates that this chemical should bioaccumulate and cause contamination in the food chain.
- BCF indicates that this chemical should bioaccumulate, and that residues could be expected in the food chain. It may be metabolized in animals; however, metabolism should be very slow.
- This chemical should bioaccumulate and cause food-chain contamination. Residues could be expected in the food chain.

Exposure

Routes of exposure could be by (1) consumption of contaminated water and food if such occurred, and (2) inhalation if the chemical were volatile; however, no data are presented herein on volatility.

Chloroethane (also Ethyl chloride)

Data

WS	5,740 mg/L	[4]
Kow	34.67	[4]
VP at 20°C	1,000 torr	[4]

Discussion

- WS indicates that this chemical should leach, run off, and be biodegraded, and it should not adsorb to soil and should not be bioaccumulated.
- Kow indicates that the chemical should not bioaccumulate, and this is in agreement with the value for WS.
- VP indicates volatility, and this could cause inhalation problems. If the chemical were phototransformed, photoproducts and the parent chemical could fall out and contaminate water and food.
- The chemical should leach and run off into aquatic environments; however, this may not be a problem because of its high volatility and potential to biodegrade. Inhalation of the volatile parent chemical and potential photoproducts should be the major concern. Contamination of the food chain is unlikely.

Exposure

Route of exposure should be inhalation of volatile residues.

Bis (2-Chloroethoxy) methane

Data

WS	81,000 mg/L	[4]
Kow	18	[4]
VP at 20°C	<0.1 torr	[4]

Discussion

- WS indicates that this chemical could leach, run off, and be biodegraded, and it should not adsorb in soil and should not be bioaccumulated.
- Kow indicates that bioaccumulation should not occur; however, if residues did get into the environment, some residues could be expected in food and water.
- VP indicates volatility, and fallout could contaminate food and water. If the chemical were phototransformed, transformation products also could contaminate food and water via fallout. Inhalation of the parent chemical and its potential photoproducts could be a problem.

Exposure

Routes of exposure could be by (1) ingestion of contaminated food and water and (2) inhalation of a volatile residue.

Bis (2-Chloroethyl) ether

Data

WS	10,200 mg/L	[4]
Kow	38	[4]
VP at 20°C	0.71 torr	[4]

Discussion

- WS indicates that this chemical could leach, run off, and be biodegraded, and it should not adsorb to soil and should not be bioaccumulated.
- Kow indicates that the chemical should not bioaccumulate, and this is supported by the value for WS.
- VP indicates volatility, and thus the potential for inhalation problems. If this chemical is phototransformed, photoproducts and the parent chemical should fall out onto water and food, causing food-chain contamination.
- This chemical could contaminate the environment by leaching, run-off, and volatility. No data have been presented to indicate if biodegradation is rapid enough to prevent residues from occurring in the environment. Inhalation of volatile residues could be a problem.

Exposure

Routes of exposure could be by (1) consumption of contaminated water and food if such occurred, and (2) inhalation; however, additional data are needed to assess problems caused by volatility.

Chloroform

Data

WS	8,200 ppm	[4]
VP	150.5 torr at 20°C	[4]

Hydrolysis	$t_{1/2}$ 3,500 years	[4]
Light adsorption	180 to 240 nm range	[4]
Pure chloroform is		
light-sensitive.		[9]

Discussion

- WS indicates leaching and runoff could occur; however, VP is so high that volatility should occur first.
- VP indicates volatility, and light adsorption indicates phototransformation.
- Hydrolysis indicates that the chemical is stable and should not be hydrolyzed in soils or in water. It could be biodegraded or phototransformed if it were to get into the aquatic environment; however, if it were not ingested, there would be no problem. It could come into direct contact with animals through volatilization.

Exposure

(1) Direct contact, (2) ingestion of contaminated water, or (3) inhalation could be the routes of exposure. Chloroform should volatilize and be phototransformed. If it got into the aquatic environment, it should not bioaccumulate, based on WS, and should volatilize from water unless it reached groundwater, where it could be stable. Inhalation should be the most direct route of exposure.

Chloromethane (also Methyl chloride)

Data

WS at 20°C	6,450 to 7,250 mg/L	[4]
Kow	8	[4]
VP at 20°C	3,756 torr	[4]

Discussion

- WS indicates that this chemical should leach, run off, and be biodegraded, and it should not be bioaccumulated and should not be adsorbed.
- Kow indicates that the chemical should not bioaccumulate, and this is in agreement with the value for WS.

- VP indicates that volatility should occur for this chemical; thus inhalation could be a problem. If it were phototransformed, photoproducts and the parent chemical could fall out and contaminate water and food.
- This chemical should be mobile in the environment and could contaminate water and food, and thus could result in contamination of the food chain. Inhalation of volatile residues could be a problem.

Exposure

Routes of exposure could be by (1) consumption of contaminated water and food if such occurred, and (2) inhalation of volatile residues.

Bis (2-Chloromethyl) ether

Data

WS	22,000 mg/L	[4]
VP at 22°C	30 torr	[4]
Kow	2.4	[4]
Hydrolyzes on contact with water	$t_{1/2}$ 10 to 40 sec	[4]

Discussion

- WS indicates that this chemical could run off, leach, and be biodegraded, but it should not bioaccumulate and should not be adsorbed to soil.
- VP indicates volatility; however, this chemical in the presence of moisture should hydrolyze in the atmosphere. Hydrolysis should prevent future exposure of the parent chemical; however, the hydrolytic products need to be studied.
- Kow indicates that this chemical should not bioaccumulate.
- It hydrolyzes with a rapid $t_{1/2}$ of 40 sec. Because the chemical volatilizes and hydrolyzes with moisture, inhalation could be a problem with effects on or in mucous membranes, lungs, and perhaps the digestive system.

Exposure

The main route of exposure could be by inhalation, which could result in problems in the mucous membranes, lungs, and digestive system. Ingestion of contaminated drinking water could be a short-term problem if consumption occurred immediately after contamination; however, this is highly unlikely because of the chemical's rapid hydrolytic rate.

Chloroneb (1,4-Dichloro-2,5-dimethoxybenzene) (also Demosan and Tersan SP)

Data

Pesticide
WS	8 ppm	[6]
Koc	1,159	[6]

Discussion

- WS indicates that this chemical should be adsorbed to soil, run off with soil, and be bioaccumulated, and it should not leach and should not be biodegraded.
- Koc indicates that this chemical should adsorb in soil and accumulate in the soil environment. Bioaccumulation could be expected, and this chemical could cause food-chain contamination. Residues should be expected in the food chain.

Exposure

Routes of exposure could be by (1) consumption of contaminated water and food if such occurred, and (2) inhalation if the chemical were volatile; however, no data are presented herein on volatility.

2-Chlorophenol (also o-Chlorophenol)

Data

Pesticide
WS at 20°C	28,500 mg/L	[4]
Kow	148	[4]
VP at 20°C	2.2 torr	[4]

Discussion

- WS indicates that this chemical could leach, run off, and be biodegraded, and it should not adsorb in soil and should not be bioaccumulated.
- Kow indicates that bioaccumulation should not occur; however, residues could be expected in the food chain.
- VP indicates volatility, and fallout could contaminate food and water. If the chemical were phototransformed, the transformation products could also contaminate food and water. Inhalation of the parent chemical and its potential photoproducts could be a problem.

Exposure

Routes of exposure could be by (1) ingestion of contaminated food and water and (2) inhalation of a volatile residue.

4-Chlorophenyl phenyl ether

Data

WS	59 mg/L	[4]
Kow	100,000	[4]
VP at 20°C	0.001 torr	[4]

Discussion

- WS indicates that this chemical could bioaccumulate, adsorb to soil, and run off with soil particles, and it should not leach and should not be biodegraded. If Kow were unknown, this prediction would be difficult, as this chemical could go either way.
- Kow indicates bioaccumulation and potential for food-chain contamination.
- VP indicates that volatility may be likely, and phototransformation could occur. Inhalation and fall out of photo products would be a problem.

Exposure

The routes of exposure could be by (1) ingestion of contaminated food or water if such contamination should occur, and (2) inhalation of volatile residues.

Chloroxuron (3-[p-(Chlorophenoxy) phenyl]-1,1-dimethlyurea) (also Tenoran)

Data

Pesticide		
WS	0.2 ppm	[6]
Koc	6,790	[6]
Soil TLC-Rf	0.09	[1]

Discussion

- WS indicates that this chemical should adsorb to soil, run off with soil, and be bioaccumulated, and it should not leach and should not be biodegraded.
- Koc indicates that this chemical should be adsorbed to soil, run off with soil, and be bioaccumulated, and it should not be leached and should not be biodegraded.
- Soil TLC indicates that this chemical should not leach and should be adsorbed to soil.
- The three pieces of data complement each other; thus the predictions should be conclusive. The data also indicate that residues could be expected in the food chain, based on the fact that the chemical may be bioaccumulated.

Exposure

Routes of exposure could be by (1) consumption of contaminated water and food if such occurred, and (2) inhalation if the chemical were volatile; however, no data are presented on volatility herein.

Chlorpropham (also Isopropyl N-(3-chlorophenyl) carbamate and CIPC)

Data

WS	88 ppm	[6]
Koc	590	[6]
Soil TLC-Rf	0.18	[1]

Discussion

- WS indicates that this chemical could go either way in relation to leaching, adsorption, runoff, biodegradation, and bioaccumulation.
- Koc indicates that this chemical should not adsorb, and that leaching could occur.
- Soil TLC indicates that this chemical should adsorb to soil to some extent and should not leach.
- The three pieces of data contradict each other; therefore, in this case, the data cannot not be used to make a prediction. Actual laboratory and perhaps field studies will be needed to discern what happens to this chemical in the environment. There will always be chemicals that do not fit the predictive mode, and one most beware of this possibility.

Exposure

Routes of exposure cannot be predicted with the data presented herein. Additional data will be needed to make conclusive predictions.

Chlorthiamid (also 2,5-Dichlorothiobenzamide and Prefix)

Data

Pesticide
WS   950 ppm   [6]
Koc   107   [6]

Discussion

- WS indicates that this chemical should adsorb to soil, run off with soil, and be bioaccumulated, and it should not leach and should not be biodegraded.
- Koc indicates that this chemical would do just the opposite of what the WS values would predict.
- Additional data would be needed to verify which piece of data could be used in predicting. Even then, actual laboratory and perhaps field data may be needed in order to make a conclusive prediction for this chemical.

Exposure

Routes of exposure cannot be predicted with the data presented above. Additional data will be needed for predictions.

Chrysene (also 1,2-Benzphenanthrene)

Data

WS at 25°C	0.002 mg/L	[4]
Kow	407,380	[4]
VP at 20°C	6.3×10^{-7} torr	[4]

Discussion

- WS indicates that this chemical could adsorb in soil, run off with soil, and be bioaccumulated, and it should not leach and should not be biodegraded.
- Kow indicates bioaccumulation, which could cause food-chain contamination. Residues could be expected in the food chain.
- VP indicates that volatility is of no concern.

Exposure

Routes of exposure could be by ingestion of contaminated food and water if such contamination should occur.

Crufomate (4-*tert*-Butyl-2-chlorophenyl methyl methylphosphoramidate) (also Ruelene)

Data

Pesticide		
WS	200 ppm	[6]
Kow	2,780	[6]

Discussion

- WS indicates that this chemical could go either way in relation to adsorption, leaching, runoff, biodegradation, and bioaccumulation.

- Kow indicates that this chemical should bioaccumulate and could cause food-chain contamination. Residues could be expected in the food chain.
- The data complement each other; thus this chemical should be expected to accumulate in the environment, and residues could be expected in the food chain.

Exposure

Routes of exposure could be by (1) consumption of contaminated water and food if such occurred, and (2) inhalation if the chemical were volatile; however, no data are presented herein on volatility.

Cyanazine (also 2-[[4-Chloro-6-(ethylamino)-s-triazin-2-yl] amino]-2-methylpropionitrile), Bladex and Payze)

Data

Pesticide		
WS	171 ppm	[6]
Koc	200	[6]
Kow	150	[6]
BCF—static water	0	[6]

Discussion

- WS indicates that this chemical could go either way in relation to leaching, runoff, adsorption, biodegradation, and bioaccumulation.
- Koc indicates that the chemical should not adsorb.
- Kow indicates that the chemical should not bioaccumulate.
- BCF indicates that the chemical should not bioaccumulate and could be metabolized in animals.
- If this chemical is released into the environment, it could be mobile, could be biodegraded, and should not bioaccumulate. There is a possibility that residues could be present in the food chain; however, there are no data presented on biodegradation and metabolism that would allow one to assess dissipation.

Exposure

Routes of exposure could be by (1) consumption of contaminated water and food if such occurred, and (2) inhalation if the chemical were volatile; however, no data are presented herein on volatility. Additional data would be needed to discern whether biodegradation and metabolism could prevent residues from being of concern.

Cycloate (also S-Ethyl cyclohexylethylthiocarbamate, Eurex, and Ro-Neet)

Data

Pesticide
WS	85 ppm	[6]
Koc	345	[6]

Discussion

- WS indicates that this chemical could go either way in regard to leaching, runoff, adsorption, biodegradation, and bioaccumulation.
- Koc indicates that the chemical could leach, run off, and be biodegraded, and it should not bioaccumulate and should not be adsorbed in soil.
- It is predicted that this chemical could be mobile and could contaminate the food chain if it were not biodegraded or metabolized rapidly enough. This prediction is based on the value for Koc and somewhat on the value for WS. Additional data are needed for a more precise prediction.

Exposure

Routes of exposure could be by (1) consumption of contaminated water and food if such occurred, and (2) inhalation if the chemical were volatile; however, no data are presented herein on volatility.

2,4-D (or 2,4-Dichlorophenoxyacetic acid)

Data

Pesticide

WS	900 ppm	[6]
Koc	20	[6]
Kow	37	[6]
BCF—static water	0	[6]
Soil TLC-Rf	0.5 to 0.69; class 4	[5]
Protein-bound		[10]

Discussion

- WS indicates that the chemical should not bioaccumulate, and that it could leach and run off.
- Koc indicates that soil adsorption should not occur, and this is supported by WS and soil TLC.
- Kow indicates that the chemical should not bioaccumulate, and food-chain contamination should not occur. This is further supported by a BCF of zero, which shows no bioaccumulation and possible metabolism in animals.
- Soil TLC indicates that leaching could occur, and this is supported by WS. The chemical could leach or run off to water and be stable in water unless degraded by microbes.

Exposure

2,4-D could leach and run off, causing contamination to the aquatic environment. Exposure could be by ingestion of contaminated water. Food-chain contamination is unlikely because of the zero BCF, which indicates metabolism by animals.

It should be noted that Yip and Ney [10] reported that 2,4-D take-up by plants and ingestion by cows resulted in residues in cow's milk, which were bound in the milk protein. The WS did not indicate this, nor could it be predicted. This is a good example of why a battery of testing and toxicological studies is needed. Since the chemical got into cow's milk, it could also get into human milk.

Dalapon (2,2-Dichloropropionic acid)

Data

Pesticide

WS	502,000 ppm	[6]
Kow	6	[6]
BCF—static water	3	[6]
Soil TLC-Rf	0.96	[1]

Discussion

- WS indicates that the chemical could leach, run off, and be biode-graded, and it should not adsorb to soil and should not be bioaccumulated.
- Kow indicates that it should not bioaccumulate.
- BCF indicates that it should not bioaccumulate, but residues may be found in the food chain.
- Soil TLC-Rf indicates that it could leach in soils.

Exposure

Routes of exposure could be by (1) drinking contaminated water, (2) ingestion of contaminated food, and (3) inhalation if the chemical were volatile; however, no data on volatility are presented herein.

DBCP (also 1,2-Dibromo-3-chloropropane, Nemagon, and Fumazone)

Data

Pesticide

WS	1,230 ppm	[6]
Koc	129	[6]
VP at 21°C	0.8 mm Hg	[8]

Discussion

- WS indicates that this chemical should leach, run off, and be biode-graded, and it should not be adsorbed and should not be bioaccumulated. This is supported by the value for Koc.

- Koc indicates that this chemical should not be adsorbed and should not be bioaccumulated, and it should leach, run off, and be biodegraded.
- VP indicates that this chemical is volatile; thus inhalation could be a problem. Volatile residues and—if the chemical were phototransformed—both parent and phototransformation products could fall out and contaminate water and the food chain.
- This chemical, if not biodegraded rapidly, could cause contamination of water and food. Inhalation would be a problem, based on its volatility.

Exposure

Routes of exposure could be by (1) inhalation of volatile residues, as well as, perhaps, phototransformation products, and (2) consumption of contaminated water and food if such occurred; however, biodegradation may be rapid enough to prevent exposure, but no data are presented on biodegradation herein.

DDD (Dichloro diphenyl dichloroethane)

Data

Pesticide
WS	0.005 ppm	[6]
Kow	1,047,000	[6]
BCF—static water	63,830	[6]

Discussion

- WS indicates that this chemical should adsorb to soil, run off with soil, and bioaccumulate, and it should not be leached and should not be biodegraded.
- Kow indicates that the chemical should bioaccumulate and could cause contamination to the food chain. This prediction is supported by the value given for BCF and WS.
- BCF indicates that this chemical should bioaccumulate and should cause food-chain contamination. It should not be metabolized to any extent in animals.
- The chemical should contaminate the food-chain if released into the environment. Residues should be expected in the food chain, and very little metabolism is expected in animals.

Exposure

Routes of exposure (1) should be by consumption of contaminated water and food if the chemical were released into the environment, and (2) could be by inhalation if it were volatile; however, no data are presented herein on volatility.

DDE

Data

Pesticide

WS	0.01 ppm	[6]
Kow	583,000	[6]
BCF—static water	27,400	[6]

Discussion

- WS indicates that this chemical should adsorb to soil, run off with soil, and be bioaccumulated, and it should not be leached and should not be biodegraded.
- Kow indicates that the chemical should bioaccumulate and could cause contamination to the food chain.
- BCF indicates that the chemical should bioaccumulate in the food chain, and little metabolism is expected in animals.
- This chemical should bioaccumuate and should cause contamination to the food chain. Very little metabolism is expected in animals. Residues should be expected in the food chain.

Exposure

Routes of exposure (1) should be by consumption of contaminated water and food if such occurred, and (2) could be by inhalation if the chemical were volatile; however, no data are presented herein on volatility.

DDT (Dichloro diphenyl trichloroethane)

Data

Pesticide

WS	0.0017 ppm	[6]
Koc	238,000	[6]
Kow	960,00	[6]
BCF—flowing water	61,600	[6]
BCF—static water	84,500	[6]

Discussion

- WS indicates that this chemical should adsorb to soil, run off with soil, and bioaccumulate, and it should not be leached and should not be biodegraded.
- Koc indicates that this chemical should adsorb to soil.
- Kow indicates that this chemical should bioaccumulate and could cause food-chain contamination, and this is supported by the BCF data.
- BCF indicates that this chemical should bioaccumulate in the food chain; thus residues could be expected.
- This chemical should be persistent in all environmental compartments, and residues could be expected throughout the food chain.

Exposure

Routes of exposure could be by (1) consumption of contaminated food and water if the chemical were released into the environment, and (2) inhalation if such occurred; however, no data are presented herein on volatility.

Dialifor (O,O-Diethyl S-(2-chloro-1-phthalimidoethyl) phosphorodithioate) (also dialifos and Torak)

Data

Pesticide

WS	0.18 ppm	[6]
Kow	49,300	[6]

Discussion

- WS indicates that this chemical should not leach and should not be biodegraded, and that it should be adsorbed and be bioaccumulated, and it can run off with soil.
- Kow indicates that this chemical should bioaccumulate in animals and could cause food-chain contamination. Residues should be expected in the food chain.
- Both pieces of data indicate bioaccumulation and the potential for food-chain contamination. Biodegradation is not expected.

Exposure

Routes of exposure could be by (1) consumption of contaminated water and food if such occurred, and (2) inhalation if the chemical were volatile; however, no data are presented herein on volatility.

Diallate (also S-(2,3-Dichloroallyl) diisopropylthiocarbamate, Avadex, and DATC)

Data

Pesticide
WS	14 ppm	[6]
Koc	1,900	[6]

Discussion

- WS indicates that this chemical should be adsorbed in soil and be bioaccumulated, and it can run off with soil; also it should not be biodegraded and should not be leached.
- Koc indicates that this chemical should adsorb in soil and could cause food-chain contamination. If this occurred, then residues should be expected in the food chain.
- This chemical could result in food-chain contamination and residues in food sources. The chemical is a carbamate derivative; therefore, it has the potential to be metabolized in animals.

Exposure

Routes of exposure could be by (1) consumption of contaminated water and food if such occurred, and (2) inhalation if the chemical were volatile; however, no data are presented herein on volatility.

Diamidofos (Phenyl N,N'-dimethylphosphorodiamidate) (also Nellite and Dowco 169)

Data

Pesticide		
WS	50,000 ppm	[6]
Koc	32	[6]
BCF—flowing water	1	[6]

Discussion

- WS indicates that this chemical should leach, run off, and be biodegraded, and it should not be adsorbed and should not be bioaccumulated. This prediction is supported by the values for Koc and BCF.
- Koc indicates that the chemical should not be adsorbed in soil.
- BCF indicates that the chemical should not bioaccumulate and could be metabolized in animals.
- If this chemical were released into the environment, it could contaminate water and the food chain; however, it should be biodegraded in soil and metabolized in animals. There is a possibility that residues could be found in the food chain, but they should dissipate in time.

Exposure

Routes of exposure could be by (1) consumption of contaminated water and food if such occurred; however, if dissipation is rapid, these residues may not be a problem, and additional data should support this prediction. Also exposure could be by (2) inhalation if the chemical were volatile; however, no data are presented herein on volatility.

Diazinon (also O,O-Diethyl O-(2-isopropyl-6-methyl-4-pyrimidinyl) phosphorothioate and Spectracide)

Data

Pesticide
WS	40 ppm	[6]
BCF—flowing water	35	[6]

Discussion

- WS indicates that this chemical could go either way in relation to leaching, runoff, adsorption, biodegradation, and bioaccumulation.
- BCF indicates the possibility of a residue, but not bioaccumulation. This chemical is an organo phosphate and is likely to cause death prior to bioaccumulation unless it is metabolized in animals. Residues could be expected in animals if the food chain is contaminated.
- Additional data would be needed to discern or predict what may happen to this chemical in soil, because the WS value in this case is not sufficient for us to make conclusive predictions.

Exposure

Routes of exposure could be by (1) drinking contaminated water, (2) ingestion of contaminated food, and (3) inhalation if the chemical were volatile; however, no data are presented herein on volatility.

Dibromochloromethane

Data

Kow	123	[4]
VP at 10.5°C	15 torr	[4]

Discussion

- Kow indicates that this chemical should not bioaccumulate. Based on the Kow value, it is predicted that this chemical could leach, run off, and be biodegraded, and it should not adsorb to soil.
- VP indicates volatility; thus inhalation could be a problem. If the chemical were phototransformed, the photoproducts and the parent chemical could fall out and contaminate water and food, thus resulting in contamination of the food chain.

- This chemical could be mobile and could contaminate water and food. Inhalation of volatile residues could be a problem. Fallout of the parent chemical and photoproducts could cause contamination of the food chain and inhalation problems.

Exposure

Routes of exposure could be by (1) consumption of contaminated water and food if such occurred, and (2) inhalation of volatile residues.

Dicamba (3,6-Dichloro-*o*-anisic acid) (also Banvel D)

Data

Pesticide		
WS	4,500 ppm	[6]
Koc	0.4	[6]
Soil TLC-Rf	0.96	[1]

Discussion

- WS indicates that this chemical should leach, run off, and be biodegraded, and it should not be adsorbed to soil and should not be bioaccumulated.
- Koc indicates that this chemical should not be adsorbed to soil and should not be bioaccumulated; thus it should be biodegraded in soils.
- Soil TLC indicates that the chemical should leach and could contaminate the aquatic environment.
- This chemical should leach and could contaminate the aquatic environment; however, it may be biodegraded prior to reaching groundwater, but no data are presented on biodegradation to support this prediction. The WS value is indicative of biodegradation, but the soil TLC is indicative of leaching. Hopefully one effect will offset the other. Additional data would help us to predict biodegradation. Residues may be expected in the food chain if biodegradation is not rapid.

Exposure

Routes of exposure could be by (1) consumption of contaminated water and food if such occurred, and (2) inhalation if the chemical were volatile; however, no data are presented herein on volatility.

Dichlobenil (2,6-Dichlorobenzonitrile) (also Casoron)

Data

Pesticide

WS	0.18 ppm	[6]
Koc	235	[6]
BCF—static water	55	[6]
Soil TLC-Rf	0.22	[1]

Discussion

- WS indicates that this chemical should adsorb to soil (this does not agree with Koc), be bioaccumulated (this does not agree with BCF), and run off with soil, and it should not leach and should not be biodegraded.
- Koc indicates that this chemical should not adsorb to soil and should leach (this is not in agreement with soil TLC).
- BCF indicates that this chemical should not bioaccumulate, and this is not in agreement with WS. BCF indicates that the chemical could be metabolized in animals; however, no data are presented on animal metabolism. The chemical, if released into the environment, could cause residues in the food chain.
- Soil TLC indicates that the chemical adsorbs to soil, which is not in agreement with Koc but agrees with WS.
- WS and soil TLC are in agreement, in that the chemical should adsorb to soil; however, Koc does not agree. BCF indicates that residues could be expected in the food chain, and they may be slowly metabolized. BCF indicates that bioaccumulation is unlikely, and this is not in agreement with WS and soil TLC. The bottom line is that residues could be expected in the food chain. Additional data would help clarify the difference in data; however, exposure can be predicted.

Exposure

Routes of exposure could be by (1) consumption of contaminated water and food if such occurred, and (2) inhalation if the chemical were volatile; however, no data are presented herein on volatility. Additional data would help us to make a more conclusive prediction.

Dichlofenthion (O-(2,6-Dichlorophenyl) O,O-diethyl phosphorothioate) (also VC-13 and Nemacide)

Data

Pesticide
WS   0.245 ppm   [6]
Kow   137,000   [6]

Discussion

• WS indicates that this chemical should adsorb to soil, run off with soil, and be bioaccumulated, and it should not be biodegraded and should not be leached. The value for Kow supports this prediction.
• Kow indicates that the chemical should bioaccumulate, and this could cause food-chain contamination.
• If this chemical were released into the environment, it could cause contamination of the food-chain.

Exposure

Routes of exposure could be by (1) consumption of contaminated water and food is such occurred, and (2) inhalation if the chemical were volatile; however, no data are presented herein on volatility.

o-Dichlorobenzene (also 1,2-Dichlorobenzene)

Data

WS at 25°C   145 mg/L   [4]
Kow   2,399   [4]
VP at 25°C   1.5 torr   [4]

Discussion

- WS in conjunction with Kow indicates that this chemical could adsorb in soil, run off with soil, and be bioaccumulated, and it should not leach and should not be biodegraded.
- Kow indicates bioaccumulation and food-chain contamination. Residues could be expected in the food chain.
- VP indicates volatility; fallout could contaminate water and food. If the chemical were phototransformed, transformation products could also fall out to contaminate food and water. Inhalation of the parent compound and its potential photoproducts could be a problem.

Exposure

Routes of exposure could be by (1) ingestion of contaminated food and water and (2) inhalation of volatile residues.

p-Dichlorobenzene

Data

Pesticide—mothballs		[6]
WS	79 ppm	[6]
Kow	2,455	[6]
VP at 20°C	0.6 torr	[4]
VP at 25°C, calculated	1.18 torr	[4]
BCF—flowing water	215	[6]

Discussion

- WS indicates that the chemical is soluble and could go either way—be mobile or nonmobile.
- Kow indicates that bioaccumulation and food-chain contamination could occur, and this is supported by a BCF of 215 in fish. Because the Kow is high, we can predict that this chemical should not leach.
- VP indicates volatility, and inhalation could be a problem.
- The chemical will cause food-chain contamination if it gets to the aquatic environment. It is volatile enough to volatilize from surfaces and from surface waters. Drinking water if contaminated, could be a problem.

Exposure

The main route of exposure should be inhalation. If phototransformation did not occur (no data are given here), then fallout to noncontaminated areas could be a problem.

4,4'-Dichlorobiphenyl

Data

WS	0.062 ppm	[6]
Kow	380,000	[6]
BCF—flowing water	215	[6]

Discussion

- WS indicates that this chemical should adsorb to soil, run off with soil, and be bioaccumulated, and it should not leach and should not be biodegraded. The Kow value supports this prediction; however, the BCF value does not. Perhaps the BCF value is an indication that the chemical is being metabolized in animals.
- Kow indicates that the chemical is lipo-soluble; thus it should bioaccumulate in animals. BCF does not support this prediction, but WS does.
- BCF indicates that the chemical should result in residues in the food chain; however, bioaccumulation should not occur. Metabolism cannot be predicted without additional data.
- This chemical should adsorb to soil and could contaminate the food chain, and thus could cause residues in food and water.

Exposure

Routes of exposure could be by (1) consumption of contaminated water and food if such occurred, and (2) inhalation if the chemical were volatile; however, no data are presented herein on volatility.

Dichlorodifluoromethane (also Freon-12 and Fluorocarbon-12)

Data

WS	280 mg/L	[4]
Kow	144.54	[4]
VP at 20°C	4,306 torr	[4]

Discussion

- WS indicates that this chemical could go either way; however, Kow indicates that bioaccumulation should not occur; therefore, it could leach, run off and biodegrade, and should not adsorb to soil and not be bioaccumulated.
- Kow indicates that bioaccumulation should not occur.
- VP indicates volatility, and this could cause inhalation problems. If the chemical were phototransformed, photoproducts and the parent chemical could fall out and contaminate the environment.
- This chemical is volatile and could present inhalation problems with the parent chemical and potential photoproducts. This type of chemical is known to damage the ozone layer in the atmosphere. It is unlikely that residues could be found in food, and it should not last long in water because of its volatility.

Exposure

The route of exposure could be inhalation of volatile residues. The major concern with this chemical should be the degradation of the ozone layer.

2,4-Dichlorophenol

Data

WS at 20°C	4,500 mg/L	[4]
Kow	562	[4]
VP at 20°C	0.12 torr	[4]

Discussion

- WS indicates that this chemical could leach, run off, and be biodegraded, and it should not adsorb to soil and should not be bioaccumulated.
- Kow indicates the possibility of bioaccumulation. Residues could be expected in the food chain.
- VP indicates volatility; fallout could cause food-chain contamination. If the chemical were phototransformed, transformation products also could fall out and cause food-chain contamination. Inhalation of the parent chemical and its photoproducts could be a problem.

Exposure

Routes of exposure could be by (1) ingestion of contaminated food and water and (2) inhalation of volatile residues.

3,6-Dichloropicolinic acid

Data

Pesticide
WS   1,000 ppm   [6]
Koc   2   [6]

Discussion

- WS indicates that this chemical should leach, run off, and be biodegraded, and it should not adsorb in soil and should not be bioaccumulated.
- Koc indicates that this chemical should not adsorb to soil and should not be bioaccumulated. WS and Koc are in agreement.
- There are insufficient data to predict whether residues could occur in the food chain. If the chemical reached the aquatic environment, then residues could be found in the food chain unless it were rapidly metabolized. This chemical could leach and run off into the aquatic environment.

Exposure

Routes of exposure could be by (1) consumption of contaminated water and food and (2) inhalation if the chemical were volatile; however, no data are presented herein on volatility.

Dichlorovos (also 2,2-Dichlorovinyl dimethyl phosphate, DDVP, and Vapona)

Data

Pesticide
WS	10,000 ppm	[6]
Kow	25	[6]

Discussion

- WS indicates that this chemical could leach, run off, and be biodegraded, and it should not adsorb to soil and should not be bioaccumulated.
- Kow indicates that bioaccumulation should not occur.

Exposure

Routes of exposure could be by (1) ingestion of contaminated food and water if such occurred, and (2) inhalation if the chemical were volatile; however, no data on volatility are presented herein.

Dieldrin (Hexachloroepoxyoctahydro-endo, exo-dimethanonaphthalene 85% and related compounds 5%)

Data

Pesticide
WS	0.022 ppm	[6]
BCF—flowing water	5,800	[6]
BCF—static water	4,420	[6]

Discussion

- WS indicates that this chemical could be adsorbed in soil, run off with soil, and be bioaccumulated, and it should not leach and should not be biodegraded.
- BCF indicates that bioaccumulation and food-chain contamination could occur. Residues could be expected in the food chain.

Exposure

Routes of exposure could be by (1) ingestion of contaminated food and water and (2) inhalation if the chemical were volatile; however, no data are presented herein on volatility. Food-chain contamination is highly likely, and so are residues of this chemical in food, based on the BCF values given.

Diethylaniline

Data

WS	670 ppm	[6]
Kow	9	[6]
BCF-static water	120	[6]

Discussion

- WS indicates that this chemical could go either way in relation to leaching, runoff, adsorption, biodegradation, and bioaccumulation. The use of Kow and BCF values in this case supports the prediction that the chemical could leach, run off, and be biodegraded, and that it should not be bioaccumulated and should not be adsorbed.
- Kow indicates that this chemical should not adsorb to soil and should not be bioaccumulated. This prediction is in agreement with WS.
- BCF indicates that the chemical should not accumulate in animals; however, residues could be expected in the food chain if contamination occurred. It is possible that metabolism could occur in animals.
- This chemical, if released into the environment, could contaminate water and the food chain. If this occurred, then residues could be expected in the food chain.

Exposure

Routes of exposure could be by (1) consumption of contaminated water and food if such occurred, and (2) inhalation if the chemical were volatile; however, no data are presented herein on volatility.

Di-2-ethylhexyl phthalate (also DEHP)

Data

Pesticide

WS	0.6 ppm	[6]
Kow	9,500	[6]
BCF—flowing water	380	[6]
BCF—static water	130	[6]

Discussion

- WS indicates that this chemical should be adsorbed in soil and be bioaccumulated and could run off with soil particles, and that it should not leach and should not be biodegraded.
- Kow indicates that this chemical could bioaccumulate.
- BCF indicates that the chemical may not bioaccumulate, perhaps because it is metabolized. The data do indicate that a chemical residue could be expected in the food chain. There is a contradiction between WS and Kow values and the BCF; however, that is acceptable because residues of the chemical can be expected in the food chain although bioaccumulation may not occur. Residues could occur in the food chain if the chemical contaminated water.

Exposure

Routes of exposure could be by (1) ingestion of contaminated food, (2) drinking contaminated water, and (3) inhalation if the chemical were volatile; however, no data on volatility are presented herein.

Diethyl phthalate (also DEP)

Data

WS	1,000 mg/L	[4]
VP	0.05 torr	[4]
Kow	26,303	[4]

Discussion

- WS indicates that the chemical could leach, run off, and be biodegraded, and that it should not bioaccumulate and should not be adsorbed in soil.
- VP indicates that the chemical could be volatile, and could fall out and contaminate food and water. If it were phototransformed, the transformation products also could fall out and contaminate food and water. Inhalation of the parent chemical and its photoproducts could be a problem.
- Kow indicates that bioaccumulation and food-chain contamination are likely. This prediction contradicts WS; however, it appears to be a valid prediction with no reason to alter it unless additional data are presented to justify a change.

Exposure

Routes of exposure could be by (1) inhalation of volatile residues and (2) ingestion of contaminated food and water.

Dimethoate (*O,O*-Dimethyl *S*-[(methylcarbanoyl) methyl]phosphorodithioate) (also Cygon)

Data

Pesticide

| WS | 25,000 ppm | [6] |
| Kow | 0.51 | [6] |

Discussion

- WS indicates that this chemical could leach, run off, and be biodegraded, and it should not bioaccumulate and should not be adsorbed in soil.

- Kow indicates that the chemical should not bioaccumulate; however, residues of the chemical may be found in the food chain.

Exposure

Routes of exposure could be by (1) ingestion of contaminated water or food if such occurred, and (2) inhalation if the chemical were volatile; however, no data are presented herein on volatility.

Dimethylnitrosamine

Data

WS	miscible	[4]
Kow	1.15	[4]
Adsorbs light	up to 400 nm	[4]
Photolysis in atmosphere	$t_{1/2}$ is 1 hour	[4]

Discussion

- WS indicates that this chemical could leach, run off, and be biodegraded, and that it should not bioaccumulate and should not be adsorbed in soil.
- Kow indicates that bioaccumulation should not occur.
- If volatile or on surfaces, this chemical could be phototransformed to other chemicals in hours. Nitrosamines are known to cause cancer; therefore, all phototransformation products are of concern. Fallout of this chemical and its photoproducts could contaminate water and the food chain. Inhalation, if the chemical is found to be volatile, is of concern for the parent chemical and its photoproducts.

Exposure

Routes of exposure (1) could be by ingestion of contaminated food or water, and (2) inhalation could be a problem, if one were continuously exposed to a volatile residue, if the chemical were found to be volatile. If volatile, the chemical should be phototransformed in the atmosphere. Phototransformation products may present an exposure problem if the formation of nitrosamines occurs.

Dimethyl phthalate (also DMP)

Data

WS	4,000 mg/L	[4]
VP at 20°C	<0.01 torr	[4]
Kow	2,630	[4]

Discussion

- WS indicates that this chemical could leach, run off, and be biodegraded, and it should not bioaccumulate and should not be adsorbed in soil.
- VP indicates that this chemical could be volatile. If it were volatile, phototransformation products could be formed and fall out to contaminate water and the food chain. Inhalation of the parent chemical and its potential photoproducts could be a problem.
- Kow indicates that bioaccumulation is likely, as is food-chain contamination. Kow contradicts WS for bioaccumulation; however, bioaccumulation is expected. A residue of DMP could be expected in water and the food chain.

Exposure

Routes of exposure could be by (1) ingestion of contaminated water and food and (2) inhalation of volatile DMP.

Dimilin (also Diflubenzuron (N-[[(4-Chlorophenyl) amino] carbonyl]-2,6-difluorobenzamide))

Data

Pesticide

WS	0.2 ppm	(6)
Koc	6,790	(6)

Discussion

- Looking at the chemical structure of Dimilin, an experienced chemist could predict some of the following:
  a. Breakdown to *p*-Chloroaníline, a pesticide, which is practically insoluble in water [8]. It is possible that this breakdown product could form azo compounds, which are known carcinogens.

b. Breakdown to difluorobenzoic acid, with WS of 1 gram in 2.3 ml of water, its solubility increasing with alkaline conditions [8].

- The breakdown products given above are only an assumption, but a good one. There could be many more breakdown products. Such assumptions cannot predict the cause of breakdown, but biodegradation and metabolism would be a good guess for it.
- WS indicates that leaching should not occur, adsorption to soil could occur, runoff with soil particles is possible, and bioaccumulation should occur unless the chemical is metabolized or biodegraded.
- Koc indicates that this chemical should adsorb to soil, and this is supported by the WS.
- Under alkaline conditions this chemical could form water-soluble chemicals, and these products could leach and contaminate sources of water.

Exposure

Exposure could be through bioaccumulation and food chain contamination. Inhalation could be a problem if the chemical were encountered in areas of aircraft applications.

Dinitramine (N^3,N^3-Diethyl-2,4-dinitro-6- (trifluoromethyl)-1,3-benzenediamine) (also Cobex)

Data

Pesticide
WS   <1 ppm   [6]
Koc   3,600   [6]

Discussion

- WS indicates that this chemical should be adsorbed to soil, should be bioaccumulated, and can run off with soil, and it should not be biodegraded and should not be leached.
- Koc indicates that the chemical should bioaccumulate and could cause contamination in the food chain. Residues could be expected in the food chain.
- This chemical, if released into the environment, could bioaccumulate and cause residues in the food chain.

Exposure

Routes of exposure could be by (1) consumption of contaminated water and food if such occurred, and (2) inhalation if the chemical were volatile; however, no data on volatility are presented herein.

4,6-Dinitro-*o*-cresol

Data

Kow   708   [4]

Discussion

- Kow indicates potential for bioaccumulation, and that residues could be expected in the food chain if contamination occurred. Kow also indicates that WS could be the cause of leaching, runoff, and biodegradation, but should not be the cause of adsorption in soil or bioaccumulation.

Exposure

Routes of exposure could be by (1) ingestion of contaminated food and water if such occurred, and (2) inhalation if the chemical were volatile; however, no date are presented herein to indicate volatility.

2,4-Dinitrophenol

Data

WS at 18°C   5,600 mg/L   [4]
Kow   34   [4]

Discussion

- WS indicates that this chemical could leach, run off, and be biodegraded, and it should not adsorb in soil and should not be bioaccumulated.
- Kow indicates that bioaccumulation should be no problem.

Exposure

Routes of exposure could be by (1) ingestion of contaminated food and water and (2) inhalation if the chemical were found to be volatile; however, no data are presented herein on volatility.

2,4-Dinitrotoluene

Data

WS at 22°C	270 mg/L	[4]
Kow	102	[4]
VP at 59°C	0.0013 torr	[4]

Discussion

- WS indicates that this chemical could leach, run off, and be biodegraded, and it should not bioaccumulate and should not be adsorbed in soil.
- Kow indicates that this chemical should not bioaccumulate, and this supports the predictions based on WS.
- VP indicates potential for volatility; fallout could contaminate food and water. If the chemical were phototransformed, transformation products could fall out, contaminating food and water. Inhalation of the parent compound and its potential photoproducts could be a problem.

Exposure

Routes of exposure could be by (1) ingestion of contaminated food and water and (2) inhalation of volatile residues.

Dinoseb (2-sec-Butyl-4,6-dinitrophenol) (also DNBP and Dyanap)

Data

Pesticide

WS	50	[6]
Koc	4	[6]
Kow	4,900	[6]

Discussion

- WS indicates that this chemical could go either way in regard to leaching, runoff, adsorption, biodegradation, and bioaccumulation.
- Koc indicates that the chemical should not adsorb in soil.
- Kow indicates that the chemical should bioaccumulate.
- The three pieces of data differ, so that it is difficult to make a prediction in regard to leaching, adsorption, runoff, and biodegradation. Based on the value given for Kow, one could predict that residues should be expected in water and food if this chemical were released into the environment. There are no data to assess how long the residues would remain in water and food. Additional data would be needed on biodegradation and metabolism in animals to discern dissipation.

Exposure

Routes of exposure could be by (1) consumption of contaminated water and food if such occurred, and (2) inhalation if the chemical were volatile; however, no data are presented herein on volatility.

Diphenylnitrosamine

Data

Kow   372   [4]

Discussion

- Kow indicates that bioaccumulation should not occur. Kow also indicates that the WS could be in a range that could cause leaching, runoff, and biodegradation, and should not cause adsorption in soil and bioaccumulation.

Exposure

Routes of exposure could be by (1) ingestion of contaminated food and water if such occurred, and (2) inhalation if the chemical were volatile; however, no data on volatility are presented herein.

Diphenyl oxide (also Phenyl ether)

Data

WS	21 ppm	[6]
Kow	15,800	[6]
BCF—flowing water	196	[6]

Discussion

- WS indicates that this chemical could go either way in regard to leaching, runoff, adsorption, bioaccumulation, and biodegradation.
- Kow indicates that the chemical should bioaccumulate and could contaminate the food chain.
- BCF indicates that the chemical should not bioaccumulate, and this is in disagreement with the data given for Kow. However, it may not be in disagreement if the chemical is metabolized in animals. Additional data would be needed to discern metabolism; however, the prediction is that the chemical could adsorb in soil and could be metabolized in animals.
- This chemical, if released into the environment, could contaminate the food chain; however, the residues could be metabolized in animals. Residues should be expected in animals, but should dissipate over time.

Exposure

Route of exposure could be by (1) consumption of contaminated water and food if such occurred, and (2) inhalation if the chemical were volatile; however, no data are presented herein on volatility.

Di-*n*-propylnitrosamine

Data

WS at 25°C	9,895 mg/L	[4]
Kow	20	[4]

Discussion

- WS indicates that this chemical could leach, run off, and be biodegraded, and it should not bioaccumulate and should not be adsorbed in soil.
- Kow indicates that bioaccumulation should not occur.

Exposure

Routes of exposure could be by (1) ingestion of contaminated food and water if that occurred, and (2) inhalation if the chemical were found to be volatile; however, no data are presented herein on volatility.

Disulfoton (also *O,O*-Diethyl *S*-[2-(ethylthio) ethyl] phosphorodithioiate, Di-Syston, Dithiodemeton, Thiodemeton, etc.)

Data

Pesticide

WS	25 ppm	[6]
Koc	1,780	[6]

Discussion

- WS indicates that this chemical should adsorb to soil, run off with soil, and be bioaccumulated, and it should not be biodegraded and should not be leached.
- Koc indicates that the chemical should bioaccumulate, and this is supported by the value for WS.
- This chemical should adsorb to soil and could bioaccumulate in the food chain. If released into the environment, it could cause residues in the food chain.

Exposure

Routes of exposure could be by (1) consumption of contaminated food if such occurred, and (2) inhalation if the chemical were volatile; however, no data are presented herein on volatility.

Diuron (3-(3,4-Dichlorophenyl)-1,1-dimethylurea)

Data

WS	42 ppm	[6]
Koc	400	[6]
Kow	94	[6]
Soil TLC-Rf	0.24	[1]

Discussion

- WS indicates the chemical could go either way with respect to leaching, runoff, biodegradation, adsorption in soil, and bioaccumulation. WS alone could not be used to predict the above; additional data would be needed.
- Koc indicates that adsorption should not occur.
- Kow indicates that bioaccumulation should not occur.
- Soil TLC indicates that leaching should not occur.
- One piece of data does not support the other. For this chemical, actual tests would be needed to discern leaching, adsorption, runoff, biodegradation, and bioaccumulation. The data presented indicate that the chemical could result in chemical residues in the food chain but not be bioaccumulated.

Exposure

Routes of exposure could be by (1) ingestion of contaminated water or food if such occurred, and (2) inhalation if the chemical were volatile; however, no data are presented on volatility herein.

DSMA (also Disodium methanearsonate)

Data

Pesticide		
WS	254,000 ppm	[6]
Koc	770	[6]

Discussion

- WS indicates that this chemical could leach, run off, and be biodegraded, and it should not adsorb to soil and should not be bioaccumulated.
- Koc indicates that this chemical should not adsorb in soil. Adsorption could be expected for arsenical compounds; however, only the experienced person would know this.
- Arsenical compounds have been found to contaminate the environment. Additional data would be needed for a more inclusive prediction.

Exposure

Routes of exposure could be by (1) ingestion of contaminated food and water if such occurred, and (2) inhalation if the chemical were volatile; however, no data are presented on volatility herein.

Dursban (also Chlorpyrifos (O,O-Diethyl O-(3,5,6-trichloro-2-pyridyl phosphorothioate))

Data

Pesticide—termite control

WS	0.3 ppm	[6]
Koc	13,600	[6]
Kow	97,700	[6]
BCF—static water	320	[6]
BCF—flowing water	450	[6]

Discussion

- WS indicates that the chemical should not leach or run off with water. Runoff with soil particles is possible.
- Koc indicates adsorption to soil, which should prevent leaching.
- Kow indicates bioaccumulation in animals.
- BCF indicates bioaccumlation and food-chain contamination.
- Based on the data, the chemical should not leach and contaminate aquatic environments. Food-chain contamination is possible if it is taken up by plants, and the plants are eaten by animals.

Exposure

It is possible that the chemical could be taken up by plants grown in soils containing it; so food-chain contamination is possible. Dursban is an organic phosphate, and its toxicity could be a problem to animals eating food or drinking water containing its residue. Inhalation could be a problem if it were volatile; however, no data are presented herein on volatility.

Endothall (7-Oxabicyclo (2.2.1) heptane-2,3-dicarboxylic acid)

Data

Pesticide		
WS	100,000 ppm	[6]
BCF—static water	0	[6]

Discussion

- WS indicates that this chemical should leach, run off, and be biodegraded, and it should not be bioaccumulated and should not be adsorbed.
- BCF indicates the chemical should not bioaccumulate, and this is supported by the value for WS.
- The value for BCF indicates that the chemical may be metabolized in animals. Short-term residues may be expected in the food chain if the chemical is released into the environment.

Exposure

Routes of exposure could be by: (1) consumption of contaminated water and food, but the exposure could be short-term if metabolism and biodegradation were rapid; and (2) inhalation if the chemical were volatile, but no data are presented herein on volatility.

Endrin (Hexachloroepoxyoctahydro-endo, endo-dimethanonaphthalene (also 1,2,3,4,10,10-hexachloro-6,7-epoxy-1,4,4a,5,6,7,8,8a-octahydro-1,4,-endo, endo-5,8-dimethanonaphthalene)

Data

Pesticide		
WS	0.024 ppm	[6]
Kow	218,000	[6]
BCF—flowing water	4,050	[6]
BCF—static water	1,360	[6]

Discussion

- WS indicates that this chemical could adsorb in soil, be bioaccumulated, and run off with soil particles, and it should not biodegrade and should not be leached.
- Kow indicates that this chemical should bioaccumulate and cause food-chain contamination.
- BCF indicates that this chemical should bioaccumulate and cause food-chain contamination. Residues of the chemical and its by-products could be expected in the food chain.

Exposure

Routes of exposure could be by (1) ingestion of contaminated water and food and (2) inhalation if the chemical were volatile; however, no data are presented on volatility herein.

EPTC (also S-Ethyl dipropylthiocarbamate and Eptam)

Data

Pesticide		
WS	365 ppm	[6]
Koc	240	[6]

Discussion

- WS indicates that this chemical could go either way in relation to leaching, runoff, adsorption, biodegradation, and bioaccumulation.

- Koc indicates that the chemical should not be adsorbed to soil.
- Using the values for Koc and WS, it is predicted that this chemical should leach, run off, and be biodegraded, and it should not be adsorbed and should not be bioaccumulated. If it were released into the environment, its residues might be found in the food chain; however, additional data would be needed to confirm this. It is also predicted that these residues should be metabolized in animals.

Exposure

Routes of exposure could be by (1) consumption of contaminated water and food and (2) inhalation if the chemical were volatile; however, no data are presented herein on volatility.

Ethion (*O,O,O',O'*-tetraethyl *S,S'*-methylene bisphosphorodithioate)

Data

Pesticide
WS   2 ppm   [6]
Koc   15,400   [6]

Discussion

- WS indicates that this chemical could adsorb to soil, be bioaccumulated, and run off with soil, and it should not leach and should not be biodegraded.
- Koc indicates that this chemical should adsorb to soil and accumulate in soil.

Exposure

Routes of exposure could be by (1) ingestion of contaminated food and water and (2) inhalation if volatility occurred; however, no data are presented herein on volatility.

Ethylbenzene

Data

WS at 20°C	152 mg/L	[4]
Kow	1,413	[4]
VP at 20°C	7 torr	[4]

Discussion

- WS indicates that this chemical could go either way in relation to leaching, runoff, adsorption, biodegradation, and bioaccumulation. Additional data would be needed to make a prediction.
- Kow indicates that bioaccumulation and food-chain contamination could occur. Residues could be expected in the food chain.
- VP indicates volatility; fallout could contaminate food and water. If the chemical were phototransformed, transformation products could fall out and contaminate food and water. Inhalation of the parent chemical and its potential photoproducts could be a problem.

Exposure

Routes of exposure could be by (1) ingestion of contaminated food and water if such occurred, and (2) inhalation of volatile residues.

Ethylene dibromide (EDB)

Data

Pesticide

WS	3,370 ppm	[6]
Koc	44	[6]

Discussion

- WS indicates that leaching, runoff, and biodegradation could occur.
- Koc indicates that food-chain contamination should not occur, nor should adsorption in soil.
- Based on the data, EDB could be mobile, and it could contaminate aquatic environments if not biodegraded.

Exposure

Contamination is possible, of aquatic environments and of air if the chemical is volatile. Exposure could be by (1) ingestion of contaminated drinking water or (2) inhalation if EDB gets into the air.

Ethylene dichloride (also 1,2-Dichloroethane)

Data

WS at 20°C	8,690 mg/L	[4]
Kow	30	[4]
VP at 20°C	61 torr	[4]

Discussion

• WS indicates that this chemical should leach, run off, and be biodegraded, and it should not adsorb to soil and should not be bioaccumulated. The value for Kow supports this prediction.
• Kow indicates that the chemical should not bioaccumulate.
• VP indicates volatility; thus inhalation could be a problem. If it were phototransformed, fallout of photoproducts and the parent chemical could contaminate water and food.
• This chemical could be mobile in the environment, and could contaminate water and food, thus resulting in contamination of the food chain. Inhalation of volatile residues could be a problem.

Exposure

Routes of exposure could be by (1) consumption of contaminated water and food if such occurred, and (2) inhalation of volatile residues.

Fenuron (3-Phenyl-1,1-dimethylurea)

Data

Pesticide		
WS	3,850 ppm	[6]
Koc	27	[6]
Kow	10	[6]

Discussion

- WS indicates that this chemical should leach, run off, and be biodegraded, and it should not be adsorbed in soil and should not be bioaccumulated.
- Koc indicates that this chemical should not adsorb in soil.
- Kow indicates that the chemical should not bioaccumulate.
- This chemical could be mobile in the environment and could cause contamination of the food chain. Residues could dissipate; however, there are no data to discern dissipation.

Exposure

Routes of exposure could be by (1) consumption of contaminated water and food if such occurred, and (2) inhalation if the chemical were volatile; however, no data are presented herein on volatility.

Fluchloralin (N-(Chloroethyl)-alpha, alpha, alpha-trifluoro-2,6-dinitro-N-propyl-p-toluidine) (also Basalin)

Data

Pesticide
WS	<1 ppm	[6]
Koc	3,600	[6]

Discussion

- WS indicates that this chemical should adsorb to soil, run off with soil, and be bioaccumulated, and it should not be biodegraded and should not be leached.
- Koc indicates that the chemical should adsorb to soil.
- The data indicate that the chemical should adsorb to soil and bioaccumulate in the food chain if released into the environment. If this occurred, residues in the food chain could be expected.

Exposure

Routes of exposure could be by (1) consumption of contaminated water and food if such occurred, and (2) inhalation if the chemical were volatile; however, no data are presented herein on volatility.

Fluoranthene

Data

WS at 25°C	0.26 mg/L	[4]
Kow	213,796	[4]
VP at 20°C	6×10^{-6} torr	[4]

Discussion

• WS indicates that this chemical could adsorb in soil, run off with soil, and be bioaccumulated, and it should not leach and should not be biodegraded.
• Kow indicates that bioaccumulation and food-chain contamination could occur. Residues could be expected in the food chain.
• VP indicates that volatility should be of no concern.

Exposure

Routes of exposure could be by ingestion of contaminated food and water if such occurred.

Fluorene

Data

WS at 25°C	1.98 mg/L	[4]
Kow	15,136	[4]
VP at 20°C	1.3×10^{-2} torr	[4]

Discussion

• WS indicates that this chemical could adsorb in soil, run off with soil, and be bioaccumulated, and it should not leach and should not be biodegraded.
• Kow indicates that bioaccumulation and food-chain contamination could occur. Residues could be expected in the food chain.
• VP indicates volatility; fallout could contaminate food and water. If the chemical were phototransformed, transformation products also could contaminate food and water. Inhalation of the parent chemical and its potential photoproducts could be a problem.

Exposure

Routes of exposure could be by (1) ingestion of contaminated food and water and (2) inhalation of volatile residues.

Formetanate (*m*-[[(Dimethylamino) methylene] amino] phenyl methylcarbamate) (also Carzol)

Data

Pesticide
WS	<1,000 ppm	[6]
BCF—static water	0	[6]

Discussion

- WS indicates that this chemical could go either way; however, since it is close to 1,000 ppm and there is a zero BCF, it is predicted that it should leach, run off, and biodegrade, and it should not adsorb in soil and should not be bioaccumulated.
- BCF indicates that there should be no bioaccumulation, and there should be metabolism in animals.
- Although there is no bioaccumulation, and there probably is metabolism in animals, it is predicated that there could be residues in the food chain until some point in time when metabolism is complete.

Exposure

Routes of exposure could be by (1) consumption of contaminated water and food if such occurred, and (2) inhalation if the chemical were volatile; however, no data are presented herein on volatility.

Glyphosate (*N*-(Phosphonomethyl) glycine)

Data

Pesticide
WS	12,000 ppm	[6]
Koc	2,640	[6]

Discussion

- WS indicates that this chemical could leach, run off and be biode-graded, and it should not bioaccumulate and should not be adsorbed to soil.
- Koc indicates that the chemical could adsorb to soil and accumulate in soil. This contradicts WS; therefore, additional data would be needed for a conclusive prediction.

Exposure

Routes of exposure could by (1) ingestion of contaminated food and water if such occurred, and (2) inhalation if the chemical were volatile; however, no data are presented on volatility herein. Additional data would be needed to clarify the contradictions.

Heptachlor (Heptachlorotetrahydro-4,7-methanoindene and related compounds)

Data

Pesticide		
WS	0.03 ppm	[6]
BCF—flowing water	17,400	[6]
BCF—static water	2,150	[6]

Discussion

- WS indicates that this chemical should adsorb to soil, run off with soil, and be bioaccumulated, and it should not be biodegraded and should not be leached.
- Both BCF values indicate that the chemical should bioaccumulate, and it could contaminate the environment, causing residues in the food chain. The BCF values indicate that little metabolism has occurred.
- This chemical could contaminate the environment if released into it. Food-chain contamination would be expected, with very little metabolism anticipated in the food chain.

Exposure

Routes of exposure could be by (1) consumption of contaminated water and food if such occurred, and (2) inhalation if the chemical were volatile; however, no data are presented herein on volatility.

Hexachlorobenzene

Data

Pesticide		
WS	0.035 ppm	[6]
Koc	3,914	[6]
Kow	168,000	[6]
BCF—flowing water	8,600	[6]
BCF—static water	290	[6]

Discussion

- WS indicates that this chemical could adsorb to soil, be bioaccumulated, and run off with soil particles, and it should not leach and should not be biodegraded.
- Koc indicates that adsorption to soil could occur, and that this chemical could accumulate in soil.
- Kow indicates that this chemical could bioaccumulate and cause food-chain contamination.
- BCF indicates that this chemical could bioaccumulate and cause food-chain contamination. Residues could be expected in the food chain.

Exposure

Routes of exposure could be by (1) ingestion of contaminated food and water if such occurred, and (2) inhalation if the chemical were volatile; however, no data are presented herein on volatility. Residues could be expected in the food chain if environmental contamination occurred.

Hexachlorobutadiene

Data

WS at 20°C	2 mg/L	[4]
Kow	5,495	[4]
VP at 20°C	0.15 torr	[4]

Discussion

- WS indicates that this chemical should be adsorbed in soil, run off with soil, and be bioaccumulated, and it should not leach and should not be biodegraded. The value for Kow supports this prediction.
- Kow indicates that the chemical should bioaccumulate.
- VP indicates that volatility could occur; thus inhalation could be a problem. If the chemical were phototransformed, photoproducts and the parent chemical could fall out and contaminate water and food.
- This chemical could contaminate the food chain because of its volatility. If food-chain contamination occurred, then residues could be expected in water and food which could result in bioaccumulation. Inhalation of volatile residues could be a problem.

Exposure

Routes of exposure could be by (1) consumption of contaminated water and food if such occurred, and (2) inhalation of volatile residues.

Hexachlorocyclopentadiene

Data

WS	0.805 mg/L	[4]
VP at 25°C	0.081 torr	[4]

Discussion

- WS indicates that this chemical could adsorb in soil, run off with soil, and be bioaccumulated, and it should not leach and should not be biodegraded.
- VP indicates that the chemical could be volatile; thus inhalation could be a problem. If it were phototransformed, photoproducts and the parent chemical could fall out and contaminate water and food.

- This chemical could contaminate water and food, resulting in contamination of the food chain. Volatile residues could present an inhalation problem.

Exposure

Routes of exposure could be by (1) consumption of contaminated water and food if such occurred, and (2) inhalation of volatile residues.

Hexachloroethane

Data

WS at 20°C	50 mg/L	[4]
Kow	2,187	[4]
VP at 20°C	0.4 torr	[4]

Discussion

- WS indicates that this chemical could go either way in regard to leaching, runoff, adsorption, biodegradation, and bioaccumulation.
- Kow indicates that the chemical should bioaccumulate. Based on the value for Kow, it is predicted that the chemical could adsorb in soil, run off with soil, and bioaccumulate, and it should not leach and should not be biodegraded. Additional data would be needed to discern biodegradation and whether metabolism in animals would occur.
- VP indicates that volatility could occur; thus inhalation could be a problem. If the chemical were phototransformed, photoproducts and the parent chemical could fall out and contaminate water and food.

Exposure

Routes of exposure could be by (1) consumption of contaminated water and food if such occurred, and (2) inhalation of volatile residues.

Imidan (also N-(Mercaptomethyl) phthalimide S-(O,O-dimethyl phosphorodithioate), Phosmet, and Prolate)

Data

Pesticide		
WS	25 ppm	[6]
Kow	677	[6]
BCF—static water	11	[6]

Discussion

- WS indicates that the chemical could go either way in regard to leaching, runoff, adsorption, biodegradation, and bioaccumulation.
- Kow also indicates that the chemical could go either way.
- BCF indicates the presence of residues, but not bioaccumulation. This chemical may have been metabolized in animals.
- The data do not agree; however, it is predicted that this chemical could leach, run off, be biodegraded, and be metabolized in animals, and this prediction is based on BCF, and the "inbetween" data for WS and Kow. Residues may be found in the food chain, but should dissipate in a short amount of time.

Exposure

Routes of exposure (1) could be by consumption of contaminated water and food if such occurred, and 2) inhalation if the chemical were volatile; however, no data are presented herein on volatility.

Ipazine (also 2-Chloro-4-(diethylamino)-6-(isopropylamine)-s-triazine)

Data

Pesticide		
WS	40 ppm	[6]
Koc	1,660	[6]
Kow	2,900	[6]

Discussion

- WS indicates that this chemical could go either way in regard to leaching, runoff, adsorption to soil, biodegradation, and bioaccumulation. Using the WS of 40 ppm, a prediction would be difficult to almost impossible to make without other data.
- Koc indicates that adsorption to soil and accumulation could occur.
- Kow indicates that bioaccumulation, food-chain contamination, and residues in the food chain could occur.

Exposure

Routes of exposure could be by (1) ingestion of contaminated food and water if such occurred, and (2) inhalation if the chemical were volatile; however, no data are presented herein on volatility.

Isocil (5-Bromo-3-isopropyl-6-methyluracil)

Data

Pesticide		
WS	2,150 ppm	[6]
Koc	130	[6]

Discussion

- WS indicates that this chemical could leach, run off, and be biodegraded, and it should not adsorb to soil and should not be bioaccumulated.
- Koc indicates that adsorption to soil should not occur.
- Bioaccumulation should not occur; however, if it reaches the aquatic environment, residues of the chemical may be found in the food chain.

Exposure

Routes of exposure could be by (1) ingestion of contaminated water and food if such occurred, and (2) inhalation if the chemical were volatile; however, no data are presented herein on volatility.

Isopropalin (2,6-Dinitro-*N,N*-dipropylcumidine)

Data

Pesticide		
WS	0.11 ppm	[6]
Koc	72,250	[6]

Discussion

- WS indicates that this chemical could adsorb in soil, be bioaccumulated, and run off with soil particles, and it should not leach and should not be biodegraded.
- Koc indicates that this chemical should adsorb to soil and accumulate in it.

Exposure

Routes of exposure could be by (1) ingestion of contaminated food and water if such occurred, and (2) inhalation if the chemical were volatile; however, no data are presented herein on volatility.

Kepone (also Decachlorooctahydro-1,3,4-metheno-2H-cyclobuta [*cd*] pentalen-2-one)

Data

Pesticide		
WS	3 ppm	[6]
BCF	8,400	[6]

Discussion

- WS indicates that kepone should not leach but could run off with soil, and it should not be biodegraded.
- BCF indicates it should bioaccumulate, cause contamination in the food chain, and be persistent in the environment.
- The author ran a computer prediction model at EPA's Athens, Georgia laboratory and predicted that kepone could persist in some aquatic soil environments over 1,000 years.

- Based on the data, bioaccumulation and food-chain contamination could occur, and persistence is long.

Exposure

Consumption of contaminated water and food could be a problem. Inhalation also could be a problem; however, no data are presented herein to indicate volatility.

Lead

Data

Is naturally occurring and has many uses
Has low water solubility   [4]
Complexes with organic material   [4]
Is not bioaccumulated   [4]
Is methylated by benthic microbes to   [4]
    form tetramethyl lead, which is volatile

Discussion

- Lead does not break down in the environment; in fact, it can be transformed to higher-molecular-weight compounds (e.g., tetramethyl lead), which are volatile.
- Lead can complex with soils; thus it can be present in clay. Clay tableware, tile roofs, etc., can contain lead. The lead can leach out of clay products, thus contaminating humans and the environment.
- Lead-based paints have contaminated humans.
- Lead solder, used to solder water pipe joints, has leached lead into drinking water.

Exposure

Routes of exposure could be by consumption of products contaminated with lead.

Leptophos (*O*-(4-Bromo-2,5-dichlorophenyl) *O*-methyl phenylphosphonothioate)

Data

Pesticide

WS	2.4 ppm	[6]
Koc	9,300	[6]
Kow	2,020,000	[6]
BCF—flowing water	750	[6]
BCF—static water	1,440	[6]

Discussion

- WS indicates that this chemical should adsorb to soil, run off with soil, and be bioaccumulated, and it should not leach and should not be biodegraded.
- Koc indicates that adsorption to soil and accumulation could occur.
- Kow indicates that bioaccumulation with potential for food-chain contamination could occur.
- BCF indicates that bioaccumulation with potential for food-chain contamination could occur.
- Residues of this chemical could be expected in the food chain.

Exposure

Routes of exposure could be by (1) ingestion of contaminated food and water if such occurred, and (2) inhalation if the chemical were volatile; however, no data are presented herein on volatility.

Lindane (Gamma isomer of benzene hexachloride)

Data

Pesticide

WS	0.15 ppm	[6]
Koc	911	[6]
BCF—flowing water	325	[6]
BCF—static water	560	[6]

Discussion

- WS indicates that the chemical could be adsorbed to soil, run off with soil, and be bioaccumulated, and it should not leach and should not be biodegraded.
- Koc indicates that the chemical should not adsorb to soil and should not be accumulated in soil, which contradicts WS.
- BCF indicates potential for bioaccumulation and food-chain contamination. BCF also indicates that residues could be expected in the food chain and that metabolism of the chemical in animals would be very slow.
- Lindane could result in residues in the food chain.

Exposure

Routes of exposure could be by (1) ingestion of contaminated food and water if such occurred, and (2) inhalation if the chemical were volatile; however, no data are presented on volatility herein.

Linuron (3-(3,4-Dichlorophenyl)-1-methoxy-1-methylurea)

Data

Pesticide		
WS	75 ppm	[6]
Koc	820	[6]
Kow	154	[6]
Soil TLC-Rf	0.17	[6]

Discussion

- WS indicates that the chemical could go either way in regard to leaching, runoff, biodegradation, adsorption, and bioaccumulation.
- Koc indicates that adsorption to soil should not occur.
- Kow indicates that bioaccumulation should not occur.
- Soil TLC indicates some adsorption, and that leaching should not occur.
- It is possible that the chemical is biodegraded and has enough adsorption to soil to prevent leaching. It is also possible for a residue to be found in the food chain, based on Kow. Additional data are needed for a more conclusive prediction.

Exposure

Routes of exposure could be by (1) ingestion of contaminated food and water if such occurred, and (2) inhalation if the chemical were volatile; however, no data are presented on volatility herein.

Malathion (O,O-Dimithyl dithiophosphate of diethyl mercaptosuccinate) (also S-(1,2-dicarbethoxyethyl) O,O-dimethyl phosophorodithioate)

Data

Pesticide

WS	145 ppm	[6]
Kow	780	[6]
BCF—static water	0	[6]

Discussion

• WS indicates that the chemical could go either way in regard to leaching, runoff, adsorption to soil, biodegradation, and bioaccumulation.
• Kow indicates that bioaccumulation should not occur.
• BCF indicates that bioaccumulation should not occur.

Exposure

Routes of exposure could be by (1) ingestion of contaminated food and water if such occurred, and (2) inhalation if the chemical were volatile; however, no data are presented herein on volatility.

Methazole (2-(3,4-Dichlorophenyl)-4-methyl-1,2,4-oxadiazolidine-3,5-dione) (also Tunic, VCS-438, Bioxone, Probe, and chlormethazole)

Data

Pesticide

WS	1.5ppm	[6]
Koc	2,620	[6]

Discussion

- WS indicates that this chemical should adsorb to soil, run off with soil, and bioaccumulate, and it should not be biodegraded and should not be leached.
- Koc indicates that the chemical could adsorb to soil or not. The value is in the in-between range; however, with the use of the value for WS, it is predicted that the chemical could bioaccumulate, adsorb to soil, and run off with soil.
- The chemical, if released into the environment, could bioaccumulate and cause contamination in the food-chain.

Exposure

Routes of exposure could be by (1) consumption of contaminated water and food if such occurred, and (2) inhalation if the chemical were volatile; however, no data are presented herein on volatility.

Methomyl (S-methyl N-[(methylcarbamoyl) oxy] thioacetimidate)

Data

Pesticide
WS	10,000 ppm	[6]
Koc	160	[6]
Kow	2	[6]

Discussion

- WS indicates leaching and runoff potential with water and possible biodegradation.
- Koc indicates that soil adsorption should not occur.
- Kow indicates that bioaccumulation and food-chain contamination should not occur, and this is further supported by WS and Koc, and vice versa.
- Based on these data, the chemical could contaminate the aquatic environment if not biodegraded, but it will not bioaccumulate.

Exposure

Exposure could be by consumption of contaminated drinking water or food if such contamination occurred. It should be noted that volatility is of concern, because of direct inhalation. This is not to be interpreted as meaning that the chemical is volatile, as no volatility data are given herein.

Methoxychlor, technical (2,2-Bis (*p*-methoxyphenyl)-1,1,1-trichloroethane 88% and related compounds 12%)

Data

Pesticide

WS	0.003 ppm	[6]
Koc	80,000	[6]
Kow	47,500	[6]
BCF—flowing water	185	[6]
BCF—static water	1,550	[6]

Discussion

- WS indicates that this chemical could adsorb in soil, run off with soil, and be bioaccumulated, and it should not leach and should not be biodegraded.
- Koc indicates that both adsorption in soil and accumulation in soil could occur.
- Kow indicates that bioaccumulation could occur and cause food-chain contamination.
- BCF indicates that bioaccumulation could occur, which would cause residues in the food chain. If metabolism occurred in animals, it would be slow, according to the BCF values.

Exposure

Routes of exposure could be by (1) ingestion of contaminated food and water if such occurred, and (2) inhalation if the chemical were volatile; however, no data on volatility are presented herein.

2-Methoxy-3,5,6-trichloropyridine

Data

WS	20.9 ppm	[6]
Koc	920	[6]
Kow	18,500	[6]

Discussion

- WS indicates that this chemical could go either way in regard to leaching, runoff, adsorption, biodegradation, and bioaccumulation.
- Koc indicates that the chemical should not adsorb and could be mobile (leach and run off).
- Kow indicates that the chemical should bioaccumulate and could cause food-chain contamination if released into the environment.
- The data are contradictory; however, values for WS and Kow can be used together to make a prediction. It is predicted that this chemical could adsorb to soil, be bioaccumulated, and run off with soil, and it should not be biodegraded and should not be leached. Kow indicates that food-chain contamination could occur; thus residues could be expected in the food chain.

Exposure

Routes of exposure could be by (1) consumption of contaminated water and food if such occurred, and (2) inhalation if the chemical were volatile; however, no data are presented herein on volatility.

9-Methylanthracene

Data

WS	0.261 ppm	[6]
Koc	65,000	[6]
Kow	117,000	[6]

Discussion

- WS indicates that this chemical should adsorb to soil, run off with soil, and bioaccumulate, and it should not be biodegraded and should not be leached. This prediction is supported by both Koc and Kow values.

- Koc indicates that the chemical should adsorb to soil.
- Kow indicates the chemical should bioaccumulate, and this could cause contamination of the food chain.
- This chemical, if released into the environment, could cause contamination of the food chain.

Exposure

Routes of exposure could be by (1) consumption of contaminated water and food if such occurred, and (2) inhalation if the chemical were volatile; however, no data are presented herein on volatility.

Methyl chloroform (also 1,1,1-Trichloroethane)

Data

WS at 20°C	480 to 4,400 mg/L	[4]
Kow	148	[4]
VP at 20°C	96 torr	[4]

Discussion

- WS has such a large range of values that it would be difficult to predict what could occur in the environment unless other data were available.
- Kow indicates that this chemical should not bioaccumulate. By using the value for Kow and the highest value for WS, the two would be in agreement. The prediction is that the chemical could leach, run off, and biodegrade, and it should not bioaccumulate and should not be adsorbed in soil.
- VP indicates volatility; thus inhalation could be a problem. If the chemical were phototransformed, photoproducts and the parent chemical could fall out and contaminate water and food.
- This chemical should be mobile in the environment; therefore, it could contaminate water and food, and thus could cause residues in the food chain. Volatile residues could be a problem for inhalation. There are no data to discern if dissipation would occur.

Exposure

Routes of exposure could be by (1) consumption of contaminated water and food if such occurred, and (2) inhalation of volatile residues.

3-Methylcholanthrene

Data

WS	0.00323 ± 0.00017 μ/ml	[3]
Kow	2,632,000 ± 701,000	[3]
Koc	1,244,046	[3]

Discussion

- WS indicates that this chemical should not leach or run off unless the runoff is with soil. It indicates bioaccumulation and accumulation.
- Kow indicates bioaccumulation and potential for food-chain contamination.
- Koc indicates adsorption to soil, which should prevent leaching.
- Based on the data, this chemical should not leach, should not be biodegraded, should bioaccumulate, should adsorb to soil, and should be persistent in the environment.

Exposure

This chemical should be kept out of the aquatic environment, as food-chain contamination is probable. If this chemical gets into soil, and crops are planted in the soil, crop uptake of the chemical is likely, as is food-chain contamination. Inhalation could be a problem; however, no data are presented herein on volatility.

Methylene chloride (also Dichloromethane)

Data

WS	13,2000 to 20,000 mg/L	[4]
Kow	17.78	[4]
VP at 20°C	362.4 torr	[4]

Discussion

- WS indicates that this chemical should leach, run off, and be biodegraded, and it should not adsorb to soil and should not be bioaccumulated.
- Kow indicates that the chemical should not bioaccumulate.
- VP indicates that this chemical is volatile, and this could cause inhalation problems. If the chemical were phototransformed, photoproducts and the parent chemical could fall out onto water and food, and thus could contaminate the food chain.
- This chemical should leach and run off into aquatic environments; however, this may not be a problem, because the chemical is very volatile. It should not last very long in any environment, because of its volatility. Biodegradation also should prevent this chemical from lasting long in the environment. The major concern should be its inhalation and phototransformation, if that occurred, and fallout of photoproducts onto water and soil. No data are presented herein on photo aspects.

Exposure

Route of exposure should be by inhalation of volatile residues.

Methyl isothiocyanate

Data

Pesticide		
WS	7,600 ppm	[6]
Koc	6	[6]

Discussion

- WS indicates that this chemical could leach, run off, and be biodegraded, and it should not adsorb in soil and should not be bioaccumulated.
- Koc indicates that adsorption in soil should not occur; therefore, accumulation in soil also should not occur.

Exposure

Routes of exposure could be by (1) ingestion of contaminated food and water if such occurred, and (2) inhalation if the chemical were volatile; however, no data are presented herein on volatility.

2-Methylnaphthalene

Data

WS	25.4 ppm	[6]
Koc	8,500	[6]
Kow	13,000	[6]

Discussion

- WS indicates that this chemical could go either way in regard to leaching, runoff, adsorption, biodegradation, and bioaccumulation.
- Koc indicates the chemical could adsorb to soil, and this is supported by the Kow value.
- Kow indicates that this chemical should bioaccumulate, and this could cause food-chain contamination.
- If this chemical were released into the environment, it could contaminate the food chain.

Exposure

Routes of exposure could be by (1) consumption of contaminated water and food if such occurred, and (2) inhalation if the chemical were volatile; however, no data are presented herein on volatility.

Methylparathion (*O,O*-Dimethyl-*O-p*-nitrophenyl phosphorothioate)

Data

Pesticide		
WS	57 ppm	[6]
Koc	9,800	[6]
Kow	82	[6]
BCF—static water	95	[6]
VP	9.5 mm Hg $\times 10^6$ at 20°C	[3]
Degrades in soil		[3]

Discussion

- WS of this chemical could pose a problem of mobility, but this is unlikely because of its Koc.
- Koc indicates that adsorption is likely.
- Kow indicates residues; however, bioaccumulation is unlikely.
- BCF indicates a residue; however, bioaccumulation is unlikely.
- VP indicates volatility, and this could be an inhalation problem.
- Biodegradation by soil microbes is assumed, as this chemical degrades in soil; however, no mention is made herein of whether soil degradation was by hydrolysis or other means.
- Based on the data, this chemical could get into air because of its VP. Bioaccumulation in the food-chain and accumulation in the soil are unlikely because this chemical is degraded in soil. Kow indicates it would be metabolized in animals.

Exposure

Two possible routes of exposure are (1) inhalation if one is in a contaminated area, and (2) ingestion only if this chemical gets into drinking water or on food that is consumed. Volatility is of concern, and fallout to noncontaminated areas is possible. No data are presented herein on phototransformation, and this possibility should be considered.

Metobromuron (3- (p-Bromophenyl) -1-methoxy-1-methylurea) (also Patoran)

Data

Pesticide		
WS	330 ppm	[6]
Koc	60	[6]

Discussion

- WS indicates that this chemical could go either way in regard to leaching, runoff, adsorption, biodegradation, and bioaccumulation.
- Koc indicates that the chemical should not adsorb.

- Based on the value for Koc and the lower end of the middle range for WS, it is predicted that this chemical could be mobile, and this could result in residues in the food chain.

Exposure

Routes of exposure could be by (1) consumption of contaminated water and food if such occurred, and (2) inhalation if the chemical were volatile; however, no data are presented herein on volatility.

Metolachlor (2-Chloro-N- (ethyl-6-methylphenyl) -N- (2-methoxy-1-methlethyl) acetamide)

Data

Pesticide		
Stable to hydrolysis	$t_{1/2}$ 200 days at 30°C	[9]
Stable to photolysis		[9]
Biodegraded by soil microbes		[9]
Has been shown to break down to over 20 chemicals		[9]
BCF	<10	[9]

Discussion

- Biodegradation by soil microbes is indicated; therefore, the chemical should be water-soluble, could leach, and could runoff.
- BFC indicates that bioaccumulation is unlikely.
- The chemical is stable to phototransformation; thus it would be stable in air, if volatile, unless it were biodegraded.

Exposure

If microbial populations do not rapidly biodegrade the chemical, then contamination of the aquatic environment is likely. Consumption of contaminated drinking water could be the route of exposure. Contaminated water could contaminate the food chain. Inhalation could be a problem; however, no data are presented herein on volatility.

Mexacarbate (4-Dimethylamino)-3, 5-xylyl methylcarbamate) (also Zectran)

Data

Pesticide		
WS	120 ppm	[6]
Kow	1,370	[6]

Discussion

- WS indicates that the chemical could go either way in regard to leaching, runoff, adsorption, biodegradation, and bioaccumulation.
- Kow indicates that the chemical should bioaccumulate.
- If this chemical were released into the environment, it could bioaccumulate and could cause contamination in the food chain. There are no data to indicate metabolism in animals; therefore, residues could be expected in the food chain.

Exposure

Routes of exposure could be by (1) consumption of contaminated water and food if such occurred, and (2) inhalation if the chemical were volatile; however, no data are presented herein on volatility.

Mirex (also Dodecachlorooctahydro-1,3,4-methano-1H-cyclobuta [cd] pentalene)

Data

Pesticide		
WS	0.6 ppm	[6]
BCF—static water	220	[6]

Discussion

- WS indicates that this chemical could adsorb in soil, run off with soil, and be bioaccumulated, and it should not leach and should not be biodegraded.
- BCF indicates potential for bioaccumulation and food-chain contamination. Residues could be expected in the food chain.

Exposure

Routes of exposure could be by (1) ingestion of contaminated water and food if such occurred, and (2) inhalation if the chemical were volatile; however, no data are presented herein on volatility.

Monolinuron (also 3-(*p*-Chlorophenyl)-1-methoxy-1-methylurea)

Data

Pesticide
WS	580 ppm	[6]
Koc	200	[6]
Kow	40	[6]

Discussion

- WS indicates that this chemical could go either way in regard to leaching, runoff, adsorption, biodegradation, and bioaccumulation.
- Koc indicates that the chemical should not adsorb to soil.
- Kow indicates that the chemical should not bioaccumulate.
- Using the three above values, it can be predicted that this chemical should leach, run off, and be biodegraded, and it should not be bioaccumulated and should not adsorb to soil. It is predicted that this chemical is mobile; thus food-chain contamination could occur. The $t_{1/2}$ for biodegradation and metabolism cannot be predicted; therefore, one may predict that residues could be in the food chain, but additional data would be needed for a more conclusive prediction.

Exposure

Routes of exposure could be by (1) consumption of contaminated water and food if such occurred, and (2) inhalation if the chemical were volatile; however, no data are presented herein on volatility. Additional data would be needed on biodegradation and metabolism to predict whether residues would be a problem in the food chain.

Monuron (3-(p-Chlorophenyl-1-dimethylurea (also Telvar)

Data

Pesticide	230 ppm	[6]
Koc	100	[6]
Kow	29	[6]

Discussion

- WS indicates that this chemical could go either way in regard to leaching, runoff, adsorption, biodegradation, and bioaccumulation.
- Koc indicates that the chemical should not adsorb.
- Kow indicates that the chemical should not bioaccumulate.
- The data are contradictory, and no two pieces agree. No prediction can be made.

Exposure

Routes of exposure cannot be predicted without additional data. However, to be on the safe side, if a prediction must be given, then it could be that consumption of contaminated water and food, as well as inhalation, could be a problem.

Naphthalene

Data

WS	31.7 ppm	[6]
Koc	1,300	[6]
Kow	2,040	[6]

Discussion

- WS indicates that this chemical is in the range of nonmobile to mobile.
- Koc indicates that adsorption to soil should occur, and solubility should not be conducive to leaching.
- Kow indicates bioaccumulation and food-chain contamination potential, and a water solubility that should not be conducive to mobility (leaching).

- Based on the data, the chemical should not leach, but it could run off With soil particles to contaminate aquatic environments. If any water is contaminated, or if food crops are grown in soils containing this chemical, then bioaccumulation and food-chain contamination are likely.

Exposure

Exposure would be by consumption of contaminated water or food if such occurred. Volatility data are not given; but this chemical could be a problem if volatile, if it were inhaled.

1-Naphthol

Data

WS	866 ± 31 µg/ml	[3]
Kow	700 ± 62	[3]
Koc	522	[3]
Soil TLC-Rf	0	[3]

Discussion

- WS indicates that leaching should not occur, but adsorption to soil should occur, runoff with soil particles could occur, and bioaccumulation is possible.
- Kow indicates that bioaccumulation could occur.
- Koc indicates adsorption to soil.
- Soil TLC-Rf indicated no leaching and showed potential for soil adsorption, bioaccumulation, and low solubility.
- Based on the data, the chemical should not leach, should not biodegrade, and should bioaccumulate.

Exposure

Exposure could be by consumption of contaminated water or food if such contamination occurred. Another route could be by inhalation if the chemical were volatile, but no data are given herein on volatility.

Nitralin (also 4-(Methylsulfonyl)-2,6-dinitro-*N*,*N*-dipropylaniline and Planavin)

Data

Pesticide		
WS	0.6 ppm	[6]
Koc	960	[6]

Discussion

- WS indicates that this chemical should adsorb to soil, run off with soil, and be bioaccumulated, and it should not be biodegraded and should not be leached.
- Koc indicates that the chemical should not be adsorbed, and this is not in agreement with the value for WS. The Koc value is very close to the value for going either way; perhaps the chemical could do that.
- In this situation, where the two pieces of data are not in agreement, and it is predicted that the Koc value could go either way, it should be predicted that the chemical could adsorb, and that, if released into the environment, it could bioaccumulate and cause food-chain contamination. Additional data would be needed to support this; but when one must make a prediction on limited data, sometimes it can be done.

Exposure

Routes of exposure could be by (1) consumption of contaminated water and food if such occurred, and (2) inhalation if the chemical were volatile; however, no data are presented herein on volatility. Additional data would be needed for a more conclusive prediction on food-chain contamination.

Nitrapyrin (2-Chloro-6-(trichloromethyl) pyridine) (also N-Serve)

Data

Pesticide		
WS	40 ppm	[6]
Koc	420	[6]
Kow	2,590	[6]

Discussion

- WS indicates that this chemical could go either way in regard to leaching, runoff, adsorption, biodegradation, and bioaccumulation.
- Koc indicates that this chemical should not be adsorbed and could be mobile.
- KOW indicates that the chemical should bioaccumulate and could cause food-chain contamination.
- WS and Kow are in close agreement; however, Koc does not agree. It is predicted that the chemical could bioaccumulate and cause food-chain contamination. Residues could be expected in the food chain. With the predictions on food-chain contamination, mobility can also be predicted. Biodegradation and metabolism cannot be predicted without additional data.

Exposure

Routes of exposure could be by (1) consumption of contaminated water and food if such occurred, and (2) inhalation if the chemical were volatile; however, no data are presented herein on volatility.

Nitrobenzene

Data

Pesticide
WS	1,780 ppm	[6]
Kow	62	[6]
BCF—static water	29	[6]

Discussion

- WS indicates that this chemical could leach, run off, and be biodegraded, and it should not adsorb in soil and should not be bioaccumulated.
- Kow indicates that bioaccumulation should not occur.
- BCF indicates that bioaccumulation should not occur.
- BCF indicates that residues could be expected in the food chain.

Exposure

Routes of exposure could be by (1) ingestion of contaminated food and water if such occurred, and (2) inhalation if the chemical were volatile; however, no data are presented herein on volatility.

2-Nitrophenol

Data

WS at 20°C	2,100 mg/L	[4]
Kow	58	[4]
VP at 49.3°C	1 torr	[4]

Discussion

- WS indicates that this chemical could leach, run off, and be biode-graded, and it should not adsorb in soil and should not be bioaccumulated.
- Kow indicates that bioaccumulation should not be a problem.
- VP indicates volatility; fallout could contaminate food and water. If the chemical were phototransformed, transformation products could contaminate food and water. Inhalation of the parent chemical and its potential photoproducts could be a problem.

Exposure

Routes of exposure could be by (1) ingestion of contaminated food and water and (2) inhalation of volatile residues.

4-Nitrophenol

Data

WS at 25°C	16,000 mg/L	[4]
Kow	81	[4]
VP at 146°C	2.2 torr	[4]

Discussion

- WS indicates that this chemical could leach, run off, and be biodegraded, and it should not adsorb in soil and should not be bioaccumulated.
- Kow indicates that bioaccumulation should not occur.
- VP indicates volatility; fallout could contaminate food and water. If the chemical were phototransformed, transformation products could fall out and contaminate food and water. Inhalation of the parent chemical and its potential photoproducts could be a problem.

Exposure

Routes of exposure could be by (1) ingestion of contaminated food and water and (2) inhalation of volatile residues.

Norflurazon (4-Chloro-5-(methylamino)-2-(alpha, alpha, alpha-trifluroro-*m*-tolyl)-3(2H)-pyridazinone) (also San 9789, Zorial, Monometflurazon, and Evital)

Data

Pesticide

WS	28 ppm	[6]
Koc	1,914	[6]
Soil TLC-Rf	0.4	[1]

Discussion

- WS indicates that this chemical could go either way in regard to leaching, runoff, adsorption, biodegradation, and bioaccumulation.
- Koc indicates that the chemical should adsorb to soil and run off with soil.
- Soil TLC indicates that leaching of this chemical should not occur.
- By using Koc and soil TLC values, the value for WS can then be used to predict that this chemical should adsorb to soil, run off with soil, and bioaccumulate, and it should not biodegrade and should not leach. All three values then will support the prediction that bioaccumulation could occur and cause food-chain contamination. Residues could be expected in the food chain if this chemical were released into the environment.

Exposure

Routes of exposure could be by (1) consumption of contaminated water and food if such occurred, and (2) inhalation if the chemical were volatile; however, no data are presented herein on volatility.

Oxadiazon (2-*tert*-Butyl-4-(2,4-dichloro-5-isopropoxyphenyl)-delta^2-1,3,4-oxadiazolin-5-one) (also Ronstar)

Data

Pesticide		
WS	0.7 ppm	[6]
Koc	3,241	[6]

Discussion

- WS indicates that this chemical should adsorb to soil, run off with soil, and be bioaccumulated, and it should not leach and should not be biodegraded.
- Koc indicates that the chemical should adsorb to soil, and this supported is by the value for WS.
- This chemical could bioaccumulate and cause food-chain contamination if released into the environment. If food-chain contamination occurred, then residues could be expected in the food chain.

Exposure

Routes of exposure could be by (1) consumption of contaminated water and food if such occurred, and (2) inhalation if the chemical were volatile; however, no data are presented herein on volatility.

Paraquat

Data

Pesticide		
WS	1,000,000 ppm	[6]
Koc	15,473	[6]
Soil TLC-Rf	0 to 0.13	[5]

Discussion

- WS indicates that this chemical could leach, run off, and be biodegraded, and it should not adsorb in soil and should not be bioaccumulated. Koc and soil TLC contradict WS predictions.
- Koc indicates that adsorption in soil should occur for this chemical; however, WS does not indicate this.
- Soil TLC indicates that this chemical should not leach and should be adsorbed in soil.
- This chemical does not adsorb to the organic matter of soils, but it is known to exchange with the soils' cation exchange capacity. Something had to cause the difference found with WS, Koc, and soil TLC, but only an experienced person or a person with extra knowledge of this chemical would know about the cation exchange. However, one could predict that adsorption was not based on adsorption to organic matter or organic carbon, as discussed above in the section on "Soil Sorption" in Chapter 2. Adsorption of some form did occur, however, as indicated by Koc and soil TLC.
- Bioaccumulation could not be predicted based on Koc. One can predict that accumulation in soil could occur, and that there is potential for residues in food and water using WS, Koc, and soil TLC.

Exposure

Routes of exposure could be by (1) ingestion of contaminated food and water if such occurred, and (2) inhalation if the chemical were volatile; however, no data are presented herein on volatility. Additional data would be needed to confirm these predictions.

Parathion (O,O-Diethyl O-*p*-nitrophenyl phosphorothioate)

Data

Pesticide		
WS	24 ppm	[6]
Koc	4,800	[6]
Kow	6,400	[6]
BCF—static water	335	[6]

Discussion

- WS indicates that this chemical could adsorb in soil, run off with soil, and be bioaccumulated, and it should not leach and should not be biodegraded.
- Koc indicates that adsorption in soil could occur, and that there is potential for accumulation in soil.
- Kow indicates that bioaccumulation could occur.
- BCF indicates that bioaccumulation and food-chain contamination could occur. Residues could be expected in the food chain.

Exposure

Routes of exposure could be by (1) ingestion of contaminated food and water if such occurred, and (2) inhalation if the chemical were volatile; however, no data are presented herein on volatility.

Pentachlorobenzene

Data

WS	0.135 ppm	[6]
Kow	154,000	[6]
BCF—flowing water	about 5,000	[6]

Discussion

- WS indicates that this chemical could adsorb in soil, run off with soil, and be bioaccumulated, and it should not leach and should not be biodegraded.
- Kow indicates that bioaccumulation could occur.
- BCF indicates that bioaccumulation and food-chain contamination could occur. Residues could be expected in the food chain.

Exposure

Routes of exposure could be by (1) ingestion of contaminated food or water if such occurred, and (2) inhalation if the chemical were volatile; however, no data are presented herein on volatility.

Pentachlorophenol

Data

Pesticide

WS	14 ppm	[6]
Koc	900	[6]
Kow	102,000	[6]
VP	0.00011 torr (calculated)	[4]
Adsorbs light	245 nm and 318 nm	[4]

Discussion

- WS is in the "either way" range.
- Koc indicates that sorption can occur.
- Kow indicates that bioaccumulation should occur; thus food-chain contamination is likely.
- VP indicates that volatility is not a problem.
- This chemical adsorbs light; therefore, phototransformation can occur in water and on surfaces.

Exposure

Exposure could be by consumption of contaminated water or food if such occurred. Inhalation could be a problem if the chemical were volatile; however, volatility is unlikely.

Phenanthrene

Data

WS	1.29 ppm	[6]
Koc	23,000	[6]
Kow	32,900	[6]
Kow	28,840	[4]
VP at 20°C	6.8×10^{-4} mmHg	[4]

Discussion

- WS indicates that this chemical should be adsorbed to soil, run off with soil, and be bioaccumulated, and it should not be biodegraded and should not be leached.

- Koc indicates that the chemical should be adsorbed to soil and could run off with soil.
- Kow indicates that the chemical should bioaccumulate and could cause food-chain contamination.
- VP indicates potential for volatility and thus inhalation problems. If the chemical were volatile, phototransformation products, if formed, along with the parent chemical, could fall out and contaminate water and the food chain.
- The four pieces of data complement each other. This chemical, if released in the environment, could bioaccumulate and cause food-chain contamination. Residues could be expected in the food chain. Volatile residues of the parent chemical and its possible photoproducts also could present an inhalation problem and a contamination problem, were they to fall out onto water and the food chain.

Exposure

Routes of exposure could be by (1) consumption of contaminated water and food if such occurred, and (2) inhalation; however, additional data are needed for a complete exposure picture.

Phenol (also Carbolic acid)

Data

WS	82,000 ppm	[6]
WS	93,000 ppm	[4]
Koc	27	[6]
VP	0.02 mm Hg	[7]
VP (supercooled liquid)	0.5293 torr	[4]
Kow	1.46	[4]

Discussion

- The WS of 82,000 to 93,000 ppm is an indication that phenol is water-soluble, and should leach into groundwater, run off into surface water, and be biodegraded.
- Koc indicates that adsorption in soil is likely.
- Based on its volatility, the chemical should volatilize from H_2O and soil surfaces.

- Kow indicates that bioaccumulation is unlikely.
- Based on its solubility, the chemical should not adsorb to soil, should be degraded by microbes, and should not bioaccumulate. This is supported by a low Koc of 27 and a low Kow of 1.46. Volatility could be a problem if the chemical were inhaled.

Exposure

Two routes of exposure are possible: (1) inhalation, because of the chemical's volatility, and (2) consumption of contaminated drinking water.

Phosalone (S-[6-Chloro-3-mercaptomethyl)-2-benzoxazolinone] O,O-diethyl phosphorodithioate) (also Zolone)

Data

Pesticide		
WS	10 ppm	[6]
Kow	20,100	[6]

Discussion

- WS indicates that this chemical should be adsorbed to soil, be bioaccumulated, and run off with soil, and it should not be biodegraded and should not be leached.
- Kow indicates that the chemical should be bioaccumulated and could cause food-chain contamination.
- The data support a prediction that the chemical should bioaccumulate and could contaminate the food chain. If this occurred, residues could be expected in the food chain.

Exposure

Routes of exposure could be by (1) consumption of contaminated water and food, and (2) inhalation if the chemical were volatile; however, no data are presented herein on volatility.

Phthalic anhydride

Data

WS	6,200 ppm	[6]
Kow	0.24	[6]
BCF—static water	0	[6]

Discussion

• WS indicates that this chemical should leach, run off, and be biodegraded, and it should not adsorb to soil and should not be bioaccumulated.
• Kow indicates that this chemical should not be bioaccumulated and should not cause food-chain contamination.
• BCF indicates that the chemical should not bioaccumulate.
• The three pieces of data support each other. This chemical should not bioaccumulate and should not contaminate the food chain. Residues should not be expected in the food chain. If the chemical were released in water, then consumption of the water could be a problem.

Exposure

Routes of exposure could be by (1) consumption of contaminated water if such occurred, and (2) inhalation if the chemical were volatile; however, no data are presented herein on volatility.

Phorate (O,O-Diethyl S-[(ethylthio) methyl] phosphorodithioate)

Data

Pesticide

WS	50 ppm	[6]
Koc	3,200	[6]

Discussion

• WS indicates that the chemical could go either way in regard to leaching, runoff, biodegradation, adsorption, and bioaccumulation.
• Koc indicates that adsorption to soil could occur. With this piece of data one could predict potential for bioaccumulation and food-chain contamination. Residues could be expected in the food chain.

Exposure

Routes of exposure could be by (1) ingestion of contaminated food and water if such occurred, and (2) inhalation if the chemical were volatile; however, no data are presented herein on volatility.

Picloram (4-Amino-3,5,6-trichloropicolinic acid) (also Tordon)

Data

Pesticide		
WS	430 ppm	[6]
Koc	17	[6]
Kow	2	[6]
BCF—static water	0.02	[6]
Soil TLC-Rf	0.84	[1]

Discussion

- WS, supported by Koc and Kow, indicates that this chemical could leach, run off, and be biodegraded, and it should not adsorb in soil and should not be bioaccumulated.
- Koc indicates that adsorption to soil should not be a problem.
- Kow indicates that bioaccumulation should not be a problem.
- BCF indicates that bioaccumulation should not be a problem.
- Soil TLC indicates that leaching is possible.
- This chemical could leach, run off, and contaminate the food chain if biodegradation were not rapid. Since no data are presented on biodegradation in soil, it is predicted that leaching and runoff could occur and result in food-chain contamination.

Exposure

Routes of exposure could be by (1) ingestion of contaminated food and water if such occurred, and (2) inhalation if the chemical were volatile; however, no data are presented herein on volatility.

Polychlorinated biphenyls (also PCB's and Aroclors)

Data

WS	0.0027 to 0.34 ppm	[4]
VP	0.0000771 to 0.00406 mm Hg	[4]
Kow	647 to 21,700	[4]
Absorbs light	280 to 320 nm range	[4]
Does not hydrolyze		[4]

Discussion

• Polychlorinated biphenyls, as a group, are benzenoid hydrocarbons with different numbers of chlorine atoms on different parts of the molecule. Aroclors, or PCB's, are mixtures of biphenyl molecules with different numbers of chlorine atoms attached to the molecule.
• WS indicates that they should not leach, should be adsorbed to soil organic matter, should not be biodegraded, and could bioaccumulate.
• Kow indicates that bioaccumulation and food-chain contamination are likely.
• VP indicates that volatility should not be a problem, and that phototransformation is likely.
• Absorption of light indicates that phototransformation is likely.
• Based on the data, these chemicals should be persistent in the environment, accumulate in soils, and bioaccumulate. Food-chain contamination is likely.

Exposure

The main route of exposure should be through consumption of contaminated drinking water and food if that occurred.

Profluralin (*N*-(Cyclopropylmethly)-alpha, alpha, alpha-trifluoro-2,6-dinitro-*N*-propyl-*p*-toluidine) (also Preqard and Tolban)

Data

Pesticide

WS	0.1 ppm	[6]
Koc	8,600	[6]

Discussion

- WS indicates that this chemical should adsorb in soil, run off with soil, and be bioaccumulated, and it should not leach and should not be biodegraded.
- Koc indicates that the chemical could go either way; however, by using the value given for WS, it is predicted that the chemical could adsorb in soil.
- If this chemical were released into the environment, it could bioaccumulate and cause contamination of the food chain.

Exposure

Routes of exposure could be by (1) consumption of contaminated water and food if such occurred, and (2) inhalation if the chemical were volatile; however, no data are presented herein on volatility.

Propham (also Isopropyl carbanilate, IPC, and prophos)

Data

Pesticide		
WS	250 ppm	[6]
Koc	51	[6]
Soil TLC-Rf	0.51	[1]

Discussion

- WS indicates that this chemical could go either way in regard to leaching, runoff, adsorption, biodegradation, and bioaccumulation.
- Koc indicates that the chemical should not be adsorbed to soil and could be mobile.
- Soil TLC indicates that the chemical could leach.
- Using all three pieces of data, one can predict that this chemical could leach, run off, and be biodegraded, and it should not be adsorbed and should not be bioaccumulated. If it were released into the environment, residues could be found in water and in the food chain, but not be bioaccumulated in animals. No data are presented on $t_{1/2}$ in soils or in animals; thus one cannot predict whether residues could be present in the food chain.

Exposure

Routes of exposure could be by (1) consumption of contaminated water and food if such occurred, and (2) inhalation if the chemical were volatile; however, no data are presented herein on volatility.

Propoxur (also o-Isopropoxyphenyl methylcarbamate, Baygon, Sendran, Tendex Suncide, Bay 9010, etc.)

Data

Pesticide		
WS	2,000 ppm	[6]
Koc	33	[6]
BCF—static water	146	[6]

Discussion

- WS indicates that this chemical should leach, run off, and be biodegraded, and it should not be adsorbed and should not be bioaccumulated.
- Kow indicates that the chemical should not bioaccumulate, and this in agreement with values given for WS and BCF.
- BCF indicates that the chemical should not bioaccumulate; however, contamination of the food chain and residues in the food chain could be expected.
- The data are in agreement. This chemical should not bioaccumulate. Contamination of the food chain could be expected if the chemical were released into the environment. If this occurred, residues could be expected in the food chain; however, metabolism should reduce the amount of residue present.

Exposure

Routes of exposure could be by (1) consumption of contaminated water and food if such occurred, and (2) inhalation if the chemical were volatile; however, no data are presented herein on volatility.

Pyrazon (5-Amino-4-chloro-2-phenyl-3(2*H*)-pyridazinone) (also Pyramin (*e*))

Data

Pesticide

WS	400 ppm	[6]
Koc	120	[6]
Soil TLC-Rf	0.44	[1]

Discussion

- WS indicates that this chemical could go either way in regard to leaching, runoff, adsorption, biodegradation, and bioaccumulation.
- Koc indicates that the chemical should not be adsorbed and could be mobile.
- Soil TLC indicates that the chemical could go either way with respect to adsorption and leaching.
- The data for this chemical are inclusive; however, the prediction is that the chemical could leach, run off, and be biodegraded. Some adsorption to soil could occur. Contamination of the food chain is possible; however, bioaccumulation should not occur. If the chemical were released into the environment, residues could be expected in the food chain, but not bioaccumulation.

Exposure

Routes of exposure could be by (1) consumption of contaminated water and food if such occurred, and (2) inhalation if the chemical were volatile; however, no data are presented herein on volatility.

Pyrene

Data

WS	0.135 ± 0.013 µg/ml	[4]
Kow	$124{,}000 \pm 11{,}000$	[4]
Koc	63,400	[4]
VP	6.85×10^{-7} torr	[4]
Photolysis absorbs solar radiation		[4]

Discussion

- WS indicates that leaching and biodegradation are unlikely; however, soil adsorption is likely.
- Koc indicates that runoff is unlikely unless with soil particles, and Kow further supports the prediction that leaching is unlikely. Koc indicates soil adsorption.
- Kow indicates that bioaccumulation and food-chain contamination may occur, and that prediction is supported by WS and Koc.
- Volatility should not be a problem.
- The chemical should not be mobile, could bioaccumulate, and could cause food-chain contamination.

Exposure

Routes of exposure could be by (1) drinking contaminated water and (2) eating contaminated plants and animals. These routes of exposure could occur if the chemical were spilled on cropland or in the aquatic environment.

Pyroxychlor (2-Chloro-6-methoxy-4-(trichloromethyl) pyridine)

Data

Pesticide		
WS	11.3 ppm	[6]
Koc	3,000	[6]
BCF—static water	239	[6]

Discussion

- WS, supported by Koc and BCF, indicates that this chemical could adsorb to soil, run off with soil, and be bioaccumulated, and it should not leach and should not be biodegraded. This could not have been predicted on WS alone because it could gone either way. However, with the use of Koc and BCF, such prediction is possible.
- Koc indicates that adsorption to soil could occur.
- BCF indicates that residues could be expedited in the food chain, as well as bioaccumulation potential.

Exposure

Routes of exposure could be by (1) ingestion of contaminated food and water if such occurred, and (2) inhalation if the chemical were volatile; however, no data are presented herein to indicate volatility.

Radon

Data

Soluble in water	[8]
Gas	[8]

Discussion

- WS indicates mobility in soil and the potential for aquatic contamination.
- Radon is a gas and can move through soil to contaminate the atmosphere.

Exposure

Two routes of exposure could be: (1) inhalation and (2) consumption of contaminated drinking water.

Ronnel (*O,O*-Dimethyl *O*-(2,4,5-trichlorophenyl) phosphorothioiate) (also Trolene, Korlan, fenchlorphos, and Viozene)

Data

Pesticide		
WS	6 ppm	[6]
Kow	46,400	[6]
VP at 25°C	8×10^{-4} mmHg	[8]

Discussion

- WS indicates that this chemical should not leach and should not be biodegraded, and it should be adsorbed to soil, be bioaccumulated, and run off with soil.
- Kow indicates that the chemical should bioaccumulate, and this is in agreement with the value for WS.

- VP indicates that the chemical could be volatile; thus inhalation could be a problem. No data are presented to indicate photo-transformation. Fallout onto water and food could be a problem.
- This chemical should bioaccumulate, and residues could be expected in the food chain. The chemical could be volatile enough to fall out onto water and food, causing food-chain contamination. Volatility also presents an inhalation problem. No data are presented to indicate whether phototransformation could occur. No data are presented on metabolism in animals so that one could discern $t_{1/2}$ in animals.

Exposure

Routes of exposure could be by (1) consumption of contaminated water and food and (2) inhalation of volatile residues.

Silvex (2-(2,4,5-Trichlorophenoxy) propionic acid)

Data

Pesticide
WS   140 ppm   [6]
Kow   2,600   [6]

Discussion

- WS, supported by Kow, indicates that this chemical could adsorb in soil, run off with soil, and be bioaccumulated, and it should not leach and should not be biodegraded. WS alone could not be used to predict this, as any of the above could go either way, based on WS. Additional data should be used for this prediction.
- Kow indicates that bioaccumulation and food-chain contamination could occur. Residues could be expected in the food chain.

Exposure

Routes of exposure could be by (1) ingestion of contaminated food and water if such occurred, and (2) inhalation if the chemical were volatile; however, no data are presented herein to indicate volatility.

Simazine (2-Chloro-4,6-bis (ethylamino)-*s*-triazine)

Data

Pesticide		
WS	3.5 ppm	[6]
Koc	135	[6]
Kow	155	[6]
BCF—flowing water	1	[6]
Soil TLC-Rf	0.45	[1]
Soil TLC-Rf	0.31 to 0.96	[5]

Discussion

- WS indicates that this chemical could leach, run off, and be biodegraded, and it should not adsorb in soil and should not be bioaccumulated.
- Koc indicates that adsorption to soil should be of no concern.
- Kow indicates that bioaccumulation should not occur; however, residues could be expected in the food chain.
- BCF indicates that bioaccumulation should not occur; however, residues could be found in the food chain.
- Soil TLC indicates that leaching could occur in some soils.

Exposure

Routes of exposure could be by (1) ingestion of contaminated food and water if such occurred, and (2) inhalation if the chemical were volatile; however, no data are presented herein to indicate volatility.

2,4,5-T (also 2,4,5-Trichlorophenoxyacetic acid)

Data

Pesticide		
WS	238 ppm	[6]
Koc	53	[6]
Kow	4	[6]
BCF—static water	25	[6]
Soil TLC-Rf	0.54	[1]

Discussion

- WS, supported With Koc and Kow, indicates that this chemical could leach, run off, and be biodegraded, and it should not adsorb to soil and should not be bioaccumulated.
- Koc indicates that adsorption in soil should be of no concern.
- BCF indicates that bioaccumulation should be of no concern; however, residues could be expected in the food chain.
- Soil TLC indicates the potential to leach.

Exposure

Routes of exposure could be by (1) ingestion of contaminated food and water if such occurred, and (2) inhalation if the chemical were volatile; however, no data are presented herein to indicate volatility.

TCE (also Trichloroethylene and Ethylene trichloride)

Data

WS at 20°C	1,100 ppm	[8]
Kow	195	[4]
VP at 20°C	57.9 torr	[4]
Does not absorb visible or near UV		[4]
Photooxidation occurs		[4]

Discussion

- WS indicates that this chemical could leach, run off, and be biodegraded, and it should not adsorb to soil and should not be bioaccumulated.
- Kow indicates that this chemical should not bioaccumulate.
- VP indicates this chemical could be volatile, and it should not be phototransformed because it does not absorb visible or UV light. Fallout onto aquatic and food environments is possible; thus contamination of the food chain could occur.
- Photooxidation has occurred, and TCE has photooxidized to form dichloroacetyl chloride and phosgene. *Note:* These breakdown prod-

ucts are mentioned because phosgene is a highly toxic gas that will produce death.

- Contamination of the food chain is possible, and residues in the food chain are possible.

Exposure

Routes of exposure could be by (1) ingestion of contaminated food and water if such occurred, and (2) inhalation of volatile residues.

Terbicil (3-*tert*-Butyl-5-chloro-6-methyl uracil)

Data

Pesticide		
WS	0.071 g/100 ml at 25°C	[2]
Soluble in water		[2]
Mobile in soil because of		
low adsorption		[2]

Discussion

- WS indicates that the chemical could leach in soil, could run off, should not be adsorbed, might be biodegraded, and should not bioaccumulate. This chemical could contaminate the aquatic environment.

Exposure

Exposure could be by consumption of contaminated food and water if such contamination occurred. If the chemical were volatile (no volatility data are given here), then inhalation would be a problem.

Terbufos (*S*-[[(1,1-Dimethylethyl) thio] methyl] O,O-diethyl phosphorodithioate) (also Counter)

Data

Pesticide		
WS	12 ppm	[6]

Discussion

- This chemical could be adsorbed in soil, run off with soil, and be bioaccumulated, and it should not leach and should not be biodegraded. The chemical, if released into the environment, could contaminate the food chain. One piece of data is not enough for a conclusive prediction; additional data are needed.

Exposure

Routes of exposure could be by (1) consumption of contaminated water and food if such occurred, and (2) inhalation if the chemical were volatile; however, no data are presented herein on volatility.

Tetracene (also Naphthacene)

Data

WS	0.0005 ppm	[6]
Koc	650,000	[6]
Kow	800,000	[6]

Discussion

- WS indicates that this chemical should adsorb to soil, run off with soil, and be bioaccumulated, and it should not leach and should not be biodegraded. This prediction is supported by both Koc and Kow values.
- Koc indicates that the chemical should adsorb to soil.
- Kow indicates that the chemical should bioaccumulate and cause contamination of the food chain.
- The data support the prediction that this chemical should bioaccumulate in the food chain; thus residues could be expected in the food chain. This chemical should be persistent in the environment, and very little biodegradation and metabolism should be expected to reduce the residues.

Exposure

Routes of exposure could be by (1) consumption of contaminated water and food if such occurred, and (2) inhalation the chemical were if volatile; however, no data are presented herein on volatility.

1,2,4,5-Tetrachlorobenzene

Data

WS	6 ppm	[6]
Kow	47,000	[6]
BCF—flowing water	4,500	[6]

Discussion

- WS indicates that this chemical should adsorb to soil, run off with soil, and be bioaccumulated, and it should not be biodegraded and should not be leached. This prediction is supported by Kow and BCF values.
- Kow indicates that the chemical should bioaccumulate and could contaminate the food chain.
- BCF indicates that the chemical should bioaccumulate, and residues should be expected in the food chain.
- The data indicate that this chemical should bioaccumulate, cause contamination in the food chain, and have residues in the food chain if it were released to the environment.

Exposure

Routes of exposure could be by (1) consumption of contaminated water and food, and (2) inhalation if the chemical were volatile; however, no data are presented herein on volatility.

1,1,2,2-Tetrachloroethane

Data

WS	2,900 mg/l	[4]
Koc	400	[4]
VP	5 torr at 25°C	[4]
Kow	363	[4]

Discussion

- WS indicates that leaching, runoff, potential for biodegradation, and soil adsorption should not occur.
- Koc indicates that adsorption should not occur.

- Kow indicates that bioaccumulation should not occur.
- VP indicates volatility. This chemical is volatile and could fall out onto noncontaminated areas.

Exposure

The major route of exposure should be by inhalation. If the chemical were phototransformed to a nontoxic substance, inhalation might not be a problem unless an immediate one. Contamination of noncontaminated areas by fallout is possible, and this may another route of exposure.

Tetrachloroethylene (also Perchloroethylene)

Data

WS	150 to 200 mg/L	[4]
Kow	758	[4]
VP at 20°C	14 torr	[4]

Discussion

- WS indicates that this chemical could go either way in regard to leaching, runoff, adsorption, biodegradation, and bioaccumulation.
- Kow indicates that bioaccumulation could occur.
- VP indicates that this chemical is volatile, and it could present inhalation problems. If it were phototransformed, fallout of photoproducts and the parent chemical could contaminate water and food.
- This chemical is volatile and could present inhalation problems with the parent and potential photoproducts. Although it may bioaccumulate, it is unlikely to do so because it should volatilize prior to bioaccumulation. The volatility should also prevent residues from occurring in food, but there could be some in water if aquatic contamination occurred.

Exposure

Routes of exposure could be by (1) inhalation of volatile residues and (2) consumption of contaminated water and food if such occurred.

Thiabendazole

Data

WS	< 50 ppm	[6]
Koc	1,720	[6]

Discussion

- WS indicates that this chemical could go either way in regard to leaching, runoff, adsorption, biodegradation, and bioaccumulation.
- Koc indicates that the chemical should adsorb to soil.
- Data are lacking that would permit a conclusive prediction. However, based on the Koc value, it is predicted that this chemical should be adsorbed to soil, should be bioaccumulated, and may cause contamination in the food chain if released into the environment.

Exposure

Routes of exposure could be by (1) consumption of contaminated water and food if such occurred, but additional data are needed to discern such contamination; and (2) inhalation if the chemical were volatile, but no data are presented herein on volatility.

Tillam (also S-Propyl butylethylthiocarbamate and Pebulate)

Data

Pesticide		
WS	60 ppm	[6]
Koc	630	[6]

Discussion

- WS indicates that this chemical could go either way in regard to leaching, runoff, adsorption, biodegradation, and bioaccumulation.
- Koc indicates that the chemical could leach, run off, and be biodegraded, and it should not be adsorbed and should not be bioaccumulated.
- Based on the value for Koc, it is predicted that this chemical could be mobile and could contaminate the food chain. Biodegradation and perhaps metabolism could be rapid enough to prevent exposure

to residues; however, no data are presented herein to indicate this. It is predicted that residues may occur in the food chain if they are not dissipated rapidly.

Exposure

Routes of exposure could be by (1) consumption of contaminated water and food if such occurred, and (2) inhalation if the chemical were volatile; however, no data are presented herein on volatility.

Toluene (also Toluol)

Data

WS at 25°C	534.8 mg/L	[4]
Kow	490	[4]
VP at 25°C	28.7 torr	[4]

Discussion

• WS, supported by Kow, indicates that this chemical could leach, run off, and be biodegraded, and it should not adsorb in soil and should not be bioaccumulated.
• Kow indicates that bioaccumulation should not occur; however, residues could be expected in the food chain.
• VP indicates volatility; fallout could contaminate food and water. If the chemical were phototransformed, transformation products also could fall out and contaminate food and water. Inhalation of the parent chemical and potential photoproducts could be a problem.

Exposure

Routes of exposure could be by (1) ingestion of contaminated food and water if such occurred, and (2) inhalation of volatile residues.

Toxaphene (Technical chlorinated camphene (67–69% chlorine))

Data

Pesticide		
WS	0.4 ppm	[6]
BCF—flowing water	26,400	[6]
BCF—static water	4,250	[6]

Discussion

- WS indicates that this chemical could adsorb in soil, run off with soil, and be bioaccumulated, and it should not leach and should not be biodegraded.
- BCF indicates that bioaccumulation and food-chain contamination could occur. Residues could be expected in the food chain.

Exposure

Routes of exposure could be by (1) ingestion of contaminated food and water if such occurred, and (2) inhalation if the chemical were volatile; however, no data are presented herein to indicate volatility.

Triallate

Data

WS	4 ppm	[6]
Koc	2,220	[6]

Discussion

- WS indicates that this chemical should adsorb to soil, runoff with soil, and be bioaccumulated, and it should not be biodegraded and should not be leached.
- Koc indicates that the chemical should adsorb to soil, and this prediction is in agreement with the value for WS.
- This chemical could bioaccumulate and cause contamination of the food-chain; however, additional data are needed for a conclusive prediction. If such contamination occurred, then residues could be expected in the food chain.

Exposure

Routes of exposure could be by (1) consumption of contaminated water and food if such contamination occurred, but additional data are needed; and (2) inhalation if the chemical were volatile, but no data are presented herein on volatility.

1,2,4-Trichlorobenzene

Data

WS	30 ppm	[6]
Kow	15,000	[6]
Kow	18,197	[4]
BCF—flowing water	491	[6]
VP at 20°C	0.42 torr	[4]

Discussion

- WS indicates that this chemical could go either way in regard to leaching, runoff, adsorption, biodegradation, and bioaccumulation.
- Kow indicates that the chemical should bioaccumulate.
- BCF indicates that the chemical should have residues in the food chain but not bioaccumulate.
- VP indicates volatility; there is the potential for an inhalation problem. If the chemical were phototransformed, the phototransformates and the parent chemical could fall out onto water and food and thus contaminate the food chain.
- The data indicate that this chemical should have residues in the food chain if the chemical were released into the environment. Bioaccumulation may be prevented by metabolism in animals; however, no data are presented so that one can discern this. Volatility could be an inhalation problem, and fallout of residues could cause food-chain contamination.

Exposure

Routes of exposure could be by (1) consumption of contaminated water and food and (2) inhalation; however, additional data are needed to permit assessment of the potential problems associated with volatility.

Trichlorofluoromethane (also Freon-11 and Fluorocarhon-11)

Data

WS	1,100 mg/L	[4]
Kow	338.84	[4]
VP at 20°C	667.4 torr	[4]

Discussion

- WS indicates that this chemical should leach, run off, and be biodegraded, and it should not be bioaccumulated and should not be adsorbed in soil.
- Kow indicates that this chemical should not bioaccumulate.
- VP indicates volatility, and this could cause inhalation problems. If it were phototransformed, fallout of photoproducts and the parent chemical could contaminate water and food.
- This chemical is volatile, and the parent chemical and potential photoproducts could cause inhalation problems. The major problem should be degradation of the ozone layer in the atmosphere. Residues that fall out onto water and food should volatilize; however, it may take longer for volatilization to occur in water than it would take on food.

Exposure

Route of exposure should be by inhalation of volatile residues. The major concern should be degradation of the ozone layer.

Trichlorofon (also Dimethyl (2,2,2-Trichloro-1-hydroxyethyl) phosphonate, Dylox, Chlorphos, and Anthon)

Data

WS	154,000 ppm	[6]
Kow	3	[6]

Discussion

- WS indicates that this chemical should leach, run off, and be biodegraded, and it should not bioaccumulate and should not be adsorbed to soil.
- Kow indicates that the chemical should not bioaccumulate, and this prediction is supported by the WS value.
- This chemical could contaminate the environment if released into it. Biodegradation, if rapid, should prevent residues from being found in the food chain; however, no data are presented herein to support this prediction.

Exposure

Routes of exposure could be by (1) consumption of contaminated water and food if such occurred, and (2) inhalation if the chemical were volatile; however, no data are presented herein on volatility.

2,4,6-Trichlorophenol

Data

WS at 25°C	800 mg/L	[4]
Kow	2,399	[4]
VP at 76.5°C	1 torr	[4]

Discussion

- WS, supported by Kow, indicates that this chemical could adsorb in soil, run off with soil, and be bioaccumulated, and it should not leach and should not be biodegraded.
- Kow indicates that bioaccumulation and food-chain contamination could occur. Residues could be expected in food and water.
- VP indicates volatility; fallout could contaminate food and water. If the chemical were phototransformed, transformation products could also fall out and contaminate food and water. Inhalation of the parent chemical and its potential photoproducts could be a problem.

Exposure

Routes of exposure could be by (1) ingestion of contaminated food and water if such occurred; also, (2) inhalation could be a problem.

Triclopyr ([(3,5,6-Trichloro-2-pyridinyl) oxy] acetic acid) (also Garlon)

Data

Pesticide		
WS	430 ppm	[6]
Koc	27	[6]
Kow	3	[6]
BCF—static water	0.02	[6]

Discussion

- WS indicates that this chemical could go either way in regard to leaching, runoff, adsorption, biodegradation, and bioaccumulation.
- Koc indicates that the chemical should not adsorb and should not be bioaccumulated; however, it could leach, run off, and be biodegraded.
- Kow indicates that the chemical should not bioaccumulate.
- BCF indicates that it should not bioaccumulate; however, residues could be expected in the food chain.
- Residues of this chemical could be expected in the food chain because the chemical should be mobile. These residues could be short-lived if the chemical were biodegraded and metabolized in animals. The BCF indicates metabolism in animals.

Exposure

Routes of exposure could be by (1) consumption of contaminated water and food if such occurred, and (2) inhalation if the chemical were volatile; however, no data are presented herein on volatility.

Triclopyr (butoxyethyl ester)

Data

Pesticide		
WS	23 ppm	[6]
Kow	12,300	[6]

Discussion

- WS indicates that this chemical could go either way in regard to leaching, run off, adsorption, biodegradation, and bioaccumulation.
- Kow indicates that this chemical should bioaccumulate and should not be biodegraded, and perhaps should not be metabolized in animals.
- This chemical, if released into the environment, should bioaccumulate and could contaminate the food chain. Residues could be expected in the food chain.

Exposure

Routes of exposure could be by (1) consumption of contaminated water and food and (2) inhalation if the chemical were volatile; however, no data are presented herein on volatility.

Triclopyr (triethylamine salt)

Data

Pesticide
WS	2,100,000 ppm	[6]
Kow	3	[6]

Discussion

- WS indicates that this chemical should leach, run off, and be biodegraded, and it should not adsorb in soil and should not be bioaccumulated.
- Kow indicates that the chemical should not bioaccumulate.
- This chemical should not be bioaccumulated in the food chain; however, this does not preclude residues in the food chain. The chemical should be biodegraded and may be metabolized in animals. Additional data are needed to support this prediction. This chemical should be mobile in the environment unless biodegradation were rapid.

Exposure

Routes of exposure could be by (1) consumption of contaminated water and food if such occurred, and (2) inhalation if the chemical were volatile; however, no data are presented herein on volatility.

Trietazine (2-chloro-4-diethylamino-6-ethylamino-s-triazine)

Data

Pesticide
WS	20 ppm	[6]
Koc	600	[6]
Kow	2,200	[6]
Soil TLC-Rf	0.36	[1]

Discussion

- WS, supported by Kow and soil TLC, indicates that this chemical could bioaccumulate and run off with soil, and it should not leach and should not be biodegraded.
- Koc indicates that there should be no adsorption to soil; however, soil TLC and WS indicate that some may occur. Because two tests indicate that some adsorption should occur, the prediction should be that adsorption should occur to some extent. Remember that a negligible amount of chemical usually can go either way.
- Kow indicates that bioaccumulation and food-chain contamination could occur. Residues could be expected in the food chain.
- Soil TLC that indicates slight movement and adsorption could occur.
- There are contradictions between the above data; however, the data indicate that bioaccumulation and food-chain contamination could occur. Residues could be expected in food and water. Some adsorption to soil and some leaching could occur.

Exposure

Routes of exposure could be by (1) ingestion of contaminated food and water if such occurred, and (2) inhalation if the chemical were volatile; however, no date are presented herein to indicate volatility.

Trifluralin (α,α,α-trifluoro-2,6-dimfro-N,N-dipropyl-*p*-toluidine)

Data

Pesticide		
WS	1×10^{-4} g/100 ml at 27°C	[2]
VP	199 mm Hg $\times 10^6$ at 29.5°C	[2]
Degraded by microbes		[2]
Phototransformed		[2]
Adsorbs on organic matter		[2]
Soil TLC-Rf	0; class 1	[5]

Discussion

- WS indicates that there should be no leaching, and runoff should occur with soil particles, as supported by adsorption and soil TLC.
- Volatility could be a problem for inhalation and fallout onto noncontaminated areas.
- Phototransformation can occur, resulting in the formation of phototransformation products.
- Adsorption to soil has occurred.
- Soil TLC indicates that leaching should not occur.
- Based on the information presented, this chemical could bioaccumulate, volatilize, and cause food-chain contamination.

Exposure

Routes of exposure could be by (1) inhalation and (2) consumption of contaminated plants and animals. If fallout of volatile residues onto noncontaminated areas occurred, then food-chain accumulation would be likely.

Urea

Data

Pesticide

WS	1,000,000 ppm	[6]
Koc	14	[6]
Kow	0.001	[6]

Discussion

- WS indicates that the chemical could leach and run off (as supported by Koc), may be biodegraded, and should not bioaccumulate (as supported by Kow).
- Koc indicates that adsorption should not occur.
- Kow indicates that bioaccumulation should not occur.
- This chemical could contaminate the aquatic environment.

Exposure

Exposure could be by (1) consumption of contaminated water if such contamination occurred. If the chemical were volatile (no volatility data are given here), (2) inhalation could be a problem.

Vinyl chloride (also Chloroethene)

Data

WS at 25°C	1.1 ppm	[4]
WS at 10°C	60 ppm	[4]
VP at 25°C	2,660 torr	[4]
Adsorbed light	at <220 nm	[4]

Discussion

- WS indicates that this chemical may or may not be mobile and may or may not bioaccumulate. Solubility appears to be a function of temperature, but this may be an incorrect assumption, depending on the controls used in testing. If the solubility was 1.1 ppm at 25°C, there might have been a loss due to VP. It was reported that at a lower temperature, 10°C, the solubility was 60 ppm, and the lower temperature could indicate a less volatile condition.
- VP indicates that the chemical could volatilize. Inhalation and fall-out could be a problem.
- Phototransformation could occur in the troposphere above the ozone layer.
- Based on the data given volatilization, leaching, and bioaccumulation could be a problem.

Exposure

Routes of exposure could be by (1) inhalation and (2) consumption of contaminated plants, animals, and water if such contamination occurred. The volatile residue could fall out and contaminate noncontaminated areas.

Warfarin (3-(α-Acetonylbenzyl)-4-hydroxycoumarin)

Data

Pesticide—rodenticide	
Drug—anticoagulant	[8]
Freely soluble in alkaline solutions	[8]
Almost insoluble in water	[8]

Discussion

- Warfarin is almost insoluble in water; thus it should not leach, and runoff should be with soil particles. It should adsorb to soil organic matter and bioaccumulate.
- Under alkaline conditions it is freely soluble in water; thus leaching and runoff are likely, but adsorption and bioaccumulation are unlikely.

Exposure

Exposure could be by consumption of contaminated water, plants, and animals. Inhalation could be a problem; however, no data are presented herein on volatility.

REFERENCES

[1] Dragun, James and Charles S. Helling, "Evaluation of Molecular Modelling Techniques to Estimate the Mobility of Organic Chemicals in Soils: II Water Solubility and the Molecular Fragment Mobility Coefficient," Procedures, 7th Annual Residue Symposium, Land Disposal of Municipal Solid and Hazardous Waste and Resource Recovery, (unpublished paper), March 1981.

[2] Environmental Protection Agency, "Adsorption Movement and Biological Degradation of Large Concentrations of Selected Pesticides in Soils," EPA-600/2-80-1 24, August 1980.

[3] Environmental Protection Agency, "Sorption Properties of Sediments and Energy-Related Pollutants," EPA-600/3-80-041, April 1980.

[4] Environmental Protection Agency, "Water-Related Environmental Fate of 129 Priority Pollutants," Contract No. 68-01-3852, Draft, January 1979.

[5] Helling, Charles S. and Benjamin C. Turner, "Pesticide Mobility: Determination by Soil Thin-Layer Chromatography," Science, vol. 162, pp. 562-563, November 1968.

[6] Kenaga, E. E. and C. A. I. Goring, "Relationship between Water Solubility, Soil Sorption, Octanol-Water Partitioning, and Concentration of Chemicals in Biota," Special Technical Publication 707, American Society for Testing Materials, 1980.

[7] Lyman, Warren J., William F. Reehl, and David H. Rosenblatt, Handbook of Chemical Property Estimation Methods, McGraw Hill Book Company, 1982.

[8] The Merck Index, Ninth Edition, Merck & Company, 1976.

[9] Ney, Ronald E., Jr., "Fate, Transport and Prediction Mode Application to Environmental Pollutants," Spring Research Symposium, James Madison University, UP, April 16, 1981.

[10] Yip, George and Ronald E. Ney, Jr., "Analysis of 2,4-D Residues in Milk and Forage," Weeds, Journal of the Weed Society of America, vol. 14, pp. 167-170, 1966.

Chapter 6
Predictive Methods for Pesticides

Fate and transport, and predictive methods can be used to determine what may happen to a chemical (pesticide) in the environment. This covers a wide gamut of different mechanisms. Environmental mechanisms could be temperature, precipitation, soil class, and physical and biological processes, etc. Prediction methods could be mathematical equations or models, computer models, etc.

Chemicals may be broken down in the environment by: light (called phototransformation—also called photodegradation); microbes (called biodegradation); plants and animals (called metabolism); and water (called hydrolysis). For convenience, *degradation* will be used exclusively to mean any form of break down. When the word *residues* is used, it can mean different chemicals or degradates of a parent chemical, plus parent chemical.

When a chemical—like a pesticide—is intentionally or accidentally placed in the environment, many mechanisms impact on the chemical's fate and transport.

In the soil environment a chemical may:
- leach through the soil
- run off the soil
- adsorb in the soil
- biodegrade in soil
- accumulate in the soil
- bioaccumulate in plants and animals or be metabolized
- volatilize
- be phototransformed
- contaminate aquatic environments

In the aquatic environment a chemical may:
- biodegrade
- photodegrade
- bioaccumulate in aquatic organisms
- volatilize
- contaminate plants, animals, and well water
- adsorb to suspended and bottom sediment

In the air a chemical may:
- photodegrade
- be inhaled
- be absorbed through skin
- fallout in non-contaminated environments

In plants a chemical may:
- be metabolized
- bioaccumulate
- be eaten by humans and other animals

In animals a chemical may:
- be metabolized
- bioaccumulate
- be eaten by humans or other animals
- be excreted.

Humans and wildlife may eat, drink, inhale, or absorb through the skin hundreds of toxic chemicals a year, and perhaps thousands over a lifetime. Based on risk analyses, exposure to one toxin may not be considered harmful; however, exposure to hundreds as an aggregate is not even considered. Exposure assessments and safety claims are based on exposure to ONE chemical. No one assesses human or wildlife safety based on exposure to hundreds of chemicals or considers the aggregate, antagonistic, and synergistic effects of this exposure. Scientists do not consider the aggregate number of chemicals we are passing to our offspring through conception and nursing.

We can take some steps to prevent our immediate exposure to harmful chemicals by: not purchasing highly toxic products; not living near toxic emissions; asking people not to smoke near us, etc. But how can you tell if you are being exposed to toxic chemicals? A good question. This is why the prediction techniques given in this book will be helpful.

This book is written to acquaint you with useful predictive techniques. Discussions will be presented on water solubility, bioaccumulation, adsorption, and leaching. Mathematical equations will be given and demonstrated on their ability to obtain data. Examples, questions, and answers are given. A list of hazardous chemicals will be given, including their fate and transport, and exposure analysis will be discussed.

When determining the degree of exposure scientists are actually determining if the risk of exposure to ONE chemical is acceptable. In judging exposure, the following processes should be considered in the assessment scheme—that is, if data are available. These processes are presented for the reader's awareness and are not all inclusive.

1. Exposure is assessed on the availability of a chemical for uptake or intake by a living organism. Conditions for this exposure are as follows:

   a. Will the exposure media be air, water, soil or food?
   b. What amount of chemical(s) will be present?
   c. What living organism(s) will be exposed?

Exposure to a toxic chemical(s) is reduced if a chemical(s) in soil is adsorbed or degraded to non-toxic chemical(s).

If a chemical is volatile, and present in the air inhaled by humans and other animals, then there may be no way to reduce the exposure. If the chemical is photodegraded to nontoxic chemicals, exposure to the toxin is reduced. Photodegradates may be just as, or more, toxic than the parent chemical. Fallout may contaminate other areas not previously contaminated.

If a chemical is present in water, dilution may be the only way to reduce the toxic effects. Even if diluted, animals may still bioaccumulate a chemical and pass it through the food chain.

Chemicals that are inhaled, absorbed through skin, or ingested may be passed to future generations via mother's milk, mother to fetus, sperm, or ovum.

2. The threat or possible adverse effects need to be identified to discern if a hazard exists. Some of the chemical effects to consider

are: is it caustic? lethal? mutagenic? teratogenic? carcinogenic? allergenic? a fire hazard? a sterilizer? an explosive? an eye or skin irritant? neurotoxic? fetotoxic? etc.? Scientists do not consider the aggregate, antagonistic, and synergistic effects that chemicals with these effects have on humans or other animals. They also do not consider co-metabolism and co-biodegradation.

3. The following should be considered when assessing adverse effects:

   a. Are the effects concentration dependent?
   b. Are the effects short- or long-termed?
   c. Are the effects irreversible?

4. What degree of exposure is considered a risk? This does not consider the aggregate, antagonistic, and synergistic effects of the hazardous chemicals available for exposure.

Why can't we have a chemical-free environment? Why can't the number of chemicals being discharged into the environment be reduced? In the interim, we can reduce the number of chemicals to which we are exposed by becoming knowledgeable.

Chapter 7
Effects of the Environment on Chemicals

General characteristics of a degradable chemical include those mechanisms by which a chemical is degraded or rendered to a less hazardous chemical(s) or detoxified. A chemical which is less hazardous is one that is degraded to a non-toxic/less hazardous chemical(s) or one whose mobility in the environment is prevented by soil adsorption. Such mechanisms can be induced by humans or by natural environmental conditions. Before proceeding it is necessary to define, although not inclusively, a few terms to avoid repetition. These terms are:

1. *Degradation* includes the mechanisms of biodegradation, chemical and physical degradation, biometabolism and metabolism, phototransformation or photodegradation, or any other means of chemical breakdown.

   a. Chemical degradation occurs to a lower molecular weight compound or is a chemical transformation to a higher molecular weight compound by hydrolysis, oxidation, reduction, or any other means or buildup. This does not imply that such degradates are less toxic than the parent chemical, although they may be.
   b. Physical degradation occurs to a lower molecular weight compound or is a chemical transformation to a higher molecular weight compound by photolysis, dissociation, or ionization.
   c. Biological degradation can also occur to higher or lower molecular weight compounds.
   d. The word *DEGRADATE(s)* will then include any of the above reactions and includes products resulting from degradation.

2. *Sorption* and/or its synonyms include sorption by organic matter, organic carbon, or cation exchange.

3. *Mobility* includes volatility, runoff, and leaching, unless otherwise specified.

A degradable chemical is a chemical that can be rendered less hazardous or detoxified, possibly by the following mechanisms:

1. Degraded in soil or water, and/or sorbed in soil,
2. Degraded chemically or physically in or on soil, and/or sorbed in soil, or
3. Degraded intentionally by human resourcefulness and not by natural environmental means, and sorbed in soil.

The intent of these mechanisms is to enable one to degrade a chemical to one that is less hazardous and unavailable for contamination or exposure to humans and the environment. This can be achieved by controlling degradation and sorption mechanisms. Such mechanisms have been practiced for years by industries such as petroleum, wood preservation, textiles, pulp and paper production, pharmaceuticals, and others. These mechanisms can be used to predict what happens to a chemical in the environment.

A chemical should be considered non-biodegradable if it is not less hazardous or detoxified because of its potential to contaminate humans and the environment, even though some degradation or sorption may have occurred. This is why environmental assessments have to be made on a case by case basis. For example: 1) rotational crops may take-up pesticide residues, and 2) ground water may be contaminated. But in either case, residue levels may be at acceptable levels.

A degradable chemical would then be one that is less hazardous to humans and the environment.

CHEMICAL MIXTURES

Chemical mixtures may well have to be organic chemicals if optimum degradation and sorption is expected. However, degradation and sorption can be inhibited due to incompatibility of the mixed chemicals. A few examples of incompatibility are listed:

1. When chemicals are in competition for soil sorptive sites and pH;

2. When one chemical stresses microbial populations needed for the degradation of another chemical;
3. When one chemical requires incorporation in the soil while the other has to be on the surface to be degraded;
4. When one chemical requires a high clay content for sorption and other does not; and
5. When one chemical requires a high moisture level for ionization while the other one cannot be under such conditions because of its high solubility and mobility.

Incompatibility within the degradable process could result in: 1) increased persistence, 2) decreased degradation, 3) increased mobility, 4) uptake of higher levels in the food chain, 5) fire, or 6) explosion. These chemicals would not be rendered less hazardous and would result in possible contamination and exposure to humans and the environment. These are a few of the reasons why compatibility of chemicals and the degradation process must be considered in order to assure optimum degradation of the chemicals and for complete exposure assessments.

HAZARDOUS ORGANICS

Hazardous organic chemicals may be degraded to non-hazardous chemicals by the following mechanisms:

1. Biologically degraded in the soil by microbes and/or enzymes, and/or metabolized by plants or animals.
2. Chemically degraded by hydrolysis or reactions with other chemicals.
3. Physically degraded by sunlight, dissociated, or ionized.

These mechanisms to degrade or detoxify chemicals can occur naturally in the environment or be induced by humans. Degradation may be induced by: 1) change in the soil pH, 2) addition of microbes, 3) addition of soil nutrients, or 4) addition of organic matter or organic carbon. The primary goal is for detoxification and sorption to prevent mobility. All of these need to be considered in risk assessments or exposure analyses.

Degradation and/or sorption needs to occur in a reasonable timeframe to avoid possible contamination and exposure to humans and the environment. If degradation is primary over sorption and it is slowed, then mobility could result as well as increased exposure. The faster the rate of degradation and sorption, the less likelihood exists for further contamination and exposure.

There are some cases where faster degradation may be impracticable. For example, if a highly toxic, volatile chemical is degraded rapidly on soil surface and by sunlight, such a mechanism may be a poor choice because of harmful short-term inhalation exposure.

IMMOBILIZATION

Immobilization of organic or inorganic chemicals that are not degraded or are very slowly degraded is needed to prevent contamination and exposure. Exposure is higher for these chemicals if they are not immobilized. Mobility can be controlled by: 1) soil incorporation to prevent volatility and photodegradation, 2) cover on the area to prevent runoff and photodegradation, or 3) sorption to soil organic matter, organic carbon, or cation exchange capacity to prevent leaching. Mobility is another means of exposure assessment.

PREVENTION OF CONTAMINATION

A chemical must be biologically, chemically, or physically degraded and/or immobilized in order to prevent or reduce environmental contamination and human exposure. Degradation and/or immobilization should occur within 90cm of soil to prevent or reduce exposure hazards. If outside this 90cm limit, degradation and/or sorption could be reduced, resulting in environmental and human exposure.

This 90cm zone is the optimum for degradation and/or sorption. It contains the highest populations of microbes needed for degradation of organic chemicals and the soil organic matter needed for soil sorption.

In some instances, a soil depth greater than the 90cm zone is needed for metals or their chelates, which are sorbed by cationic exchange capacity of high clay content.

HIGHER MOLECULAR WEIGHTS COMPOUNDS

Organics and/or inorganics can be degraded to higher molecular weight chemicals. This may be advantageous for the following reasons: 1) decreased mobility, 2) decreased water solubility, 3) increased sorptive capacity, or 4) decreased crop uptake, thus decreased exposure.

Chemicals of this type need to be degraded naturally or induced because they are usually: 1) not degraded or slowly degraded on their own, 2) highly mobile, and 3) accumulative in the food chain. Inorganic chemicals, for example: 1) chromium is reduced to insoluble oxides and hydroxides and is thus made unavailable for food chain uptake, and 2) at pH of about 6.5 cadmium, lead, zinc, copper, and sodium are precipitated as higher molecular weight metallic compounds and may be less hazardous in that form. Of course, some of these chemical reactions are reversible, so care must be taken when preforming an exposure assessment.

STUDIES NEEDED FOR ASSESSMENTS

Experimental tests are needed to assure that degradation and sorption mechanisms are working to detoxify a chemical.

Testing or the use of predictive techniques are needed on physical and chemical properties, fate (including degradation, persistence and accumulation), and transport (mobility or immobilization).

It is now time to learn about experimental tests and predictive needs. Researchers have demonstrated that the use of radioisotopic (labeled organic chemical(s)) chemical techniques are much more efficient in discerning degradation, degradate identity, plant and animal metabolism, accumulation, and mobility. There are predictive techniques which can be used to obtain data when data is limited to nonexistent. However, these predictive techniques are just such and are limited to exactness. They can be mathematical equations (which is what this book is about) or advanced chemical structure and computer modelling techniques.

TYPE LABORATORY TEST NEEDED

Predictive techniques will be discussed in chapters to come. Below are general discussions to inform you on the testing needed to perform exposure assessments. However, when there is no data or limited data then predictive techniques should be used to obtain information so one can achieve some indication of exposure.

Solubility

Solubility of a chemical in water is one of the most important physical parameters because of its multi use. Solubility can be used to predict sorption in soil, mobility, accumulation in animals, and for analytical methodology. Solubility is defined as the ability of a substance to blend uniformly with another. For the purpose of this text, we only will consider water solubility with another chemical.

Water solubility is the concentration of a compound that is in equilibrium in a saturated solution at a given temperature. Miscible is often used as the term for liquids and gases. There are standard procedures to measure water solubility.

Vapor pressure (VP)

Vapor pressure values are used as an indicator of a chemical ability to volatilize, which could increase exposure by inhalation. VP and solubility can be used to calculate rates of evaporation of dissolved organics from water. VP is the characteristic of a chemical at any given temperature of a vapor in equilibrium with its liquid or solid form.

There are standard methods to determine VP. Chemicals with a low VP, high sorptive capacity, high water solubility, or high boiling points (BP) are less likely to volatilize from soil if incorporated to at least 13cm. Chemicals with a high VP, low sorptive capacity, low water solubility, and low BP are likely to volatilize from soil in the first 13cm.

Ultraviolet adsorption (UV) & photodegradation

The ultraviolet adsorption spectra of a chemical in an aqueous solution is needed to predict if a chemical can be photodegraded. If a chemical can absorb UV light in the range of 100 to 390 A, it can be photodegraded. There are standard methods to discern UV wave lengths.

Adsorption and desorption

Adsorption and desorption, like solubility, can be used to predict mobility for a chemical. Sorption occurs in soil's organic matter or organic carbon, and in some cases to soil's cation exchange capacity.

Sorption can be measured in the laboratory by standard methods or predicted. Lab tests are expensive and if sorption to many soils are needed the cost goes higher. Predictive techniques are an excellent tool to find sorption or desorption on many soils.

Octanol water (Kow)

Octanol water partition coefficient is a measure of the affinity of a given chemical to the fatty tissue of an animal. It can be measured in the lab or predicted, as will be done later.

Leaching

Leaching is a mobility mechanism that is of major concern because of ground water contamination. Such contamination can result in the future contamination of the food chain, wells, etc. The ability of a chemical to leach through soil is a measure of its water solubility.

Have you noticed how many tests are based on water solubility? This is one reason that data on water solubility should always be chemically determined. Standard methods are available to carry out such testing. However, if actual chemical data is unavailable then predictions can be made.

Hydrolysis

Hydrolytic process of a chemical is important to determine if a chemical can be hydrolyzed by the pH occurring naturally in soil or water, or by the addition of an acid or base to soils. Hydrolysis is a chemical reaction in which water reacts with another substance or substances to produce one or more new substances.

There are standard methods to determine if chemicals hydrolyze and their rate of hydrolysis. Under natural environmental conditions, hydrolysis is not likely to occur at a pH <5, and >9, because these pHs are not likely to occur. Testing, therefore, should be in the range of pH >5 but <9; perhaps at 5, 7 and 9 would be best.

Microbial degradation

Degradation of an organic chemical by soil microbes (biodegradation) could detoxify the chemical. Microbial degradation can be affected by mixed chemicals. That is, one chemical could kill the microdes, thus preventing biodegradation of the other chemicals. Therefore, testing should be done on different mixtures. If a chemical is water soluble then it likely can be degraded by microbes.

There are standard methods to discern microbial degradation.

Predictive techniques

Predictive techniques are a useful tool when there is no or little data existing on a chemical, and a hazard assessment is needed because a chemical has been released into the environment. Such techniques can be mathematical, computer models, molecular fragmentations, or others.

Some basic criteria

In determining fate and transport of a chemical in the environment it is necessary to know what chemical(s) occur and how much is present. Such information is needed to determine if the chemical(s) will be degraded or immobilized to non-hazardous chemicals. The reason is that high amounts of some chemicals can kill the microbial populations and loverload the soil's adsorptive capacity. This may prevent detoxification or breakdown, increased mobility, and higher potential for exposure.

Acceptable degradation may be, for example:

1. Degradation by microbes should be known through the first half life (T½), but not less than 30 days or longer than 6 months. The 30 day period will determine if degradation occurs and the 6 months will determine the persistence of a chemical; or
2. Hydrolysis and photodegradation should be known through a chemical's first T½ or 30 days. The pH would be hard to maintain after 30 days.

Chapter 8
Leaching in Soil

Chemical leaching is defined as the movement of a chemical through the soil profile by the percolation of water. Of course, solvents other than water may move a chemical through soil. Water is the universal solvent and occurs naturally, and is the solvent referred to hereafter, unless otherwise specified (as in the case of a solvent spill).

The extent to which a chemical moves through soil is controlled by the amount and speed of percolation, Koc, WS, biodegradation and other chemical and physical processes which are not considered (i.e., cation exchange capacity, etc.). Leaching has been referred to as attenuation, movement, transport, mobility, elution, migration, etc. Most predictive models (mathematical, computer, etc.) refer to transport and leaching.

There are many tests that can be used to determine leaching, such as column leaching, actual field leaching, soil thin layer chromatography (soil TLC), high pressure liquid chromatography (HPLC), etc. Our discussions will use existing data to make predictions on leaching and will focus on mathematical equations.

Transport (movement) of a chemical from a spill, pesticide application, or disposal may result in contamination of the environment. Transport can be achieved by volatility, run off, leaching, or by plants and animals. There are numerous mechanisms that can hinder transport, such as rate of percolation, volume of precipitation, sorption, biodegradation, soil organic matter, etc.

HPLC

Helling and Turner (2) devised a method to determine chemical transport through soil. The procedure is called soil thin layer chromatography (soil TLC). Soil is coated on a glass plate, a known amount of a chemical is placed on the soil, and the plate is immersed in water up to the point where the chemical is placed. The chemical may be moved vertically by the eluting water to a place on the soil and no further; this is called retardation factor (Rf).

There are five Rf values (five classes) described by Helling and Turner (2) and three given in Part I. Rf is a measurement from a base line to a point when transport terminates on the soil TLC.

A researcher can also study biodegradation potential by using the soil TLC procedure. Let the chemical remain on the soil plate, let's say for 0, 15, 30, 60, and 90 days prior to leaching with water. If the chemical biodegrades, analytical testing will discern this. In all cases you will have to analyze for the presence or absence of a chemical residue, and that will be in all the five Rf areas.

Rf may also be calculated and may be used to collaborate Koc, Kow, and WS. The following equation reported by Haymaker [2] is another way to determine Rf.

$$Rf = \frac{1}{1 + (Koc)\,(\%\ OC/100)\,(ds)\,(1/\theta^{2/3} - 1)} \quad EQ.Rf\ 8\text{-}1$$

Looks difficult; however, remember we have discussed Koc, and Kd = % OC/100. Koc is the soil adsorption coefficient value either determined or predicted. Kd is the % organic carbon (OC) divided by 100. The ds is the density of soil solids expressed in g/cc, and this may be determined or looked up. θ is the pore fraction of soil and this may be determined or looked up. Sometimes OC is not available, but the organic matter (OM) is. When OM is present, then % OC = % OM divided by 100.

Another predictive tool has been presented and can be used to confirm other predictions. Now let us put it to the test and work some problems.

EXERCISE

Solve the Rf for each of the following chemicals, given θ as 0.75 and ds as 1.5 g/cc. Use two different % OC/ 100; 1% OC and 3% OC. By using two different OC percentages you will be able to visualize how OC affects the results. Evaluate fate and transport, and exposure analyses; Table 10-1, Chapter 10 will be needed to help in the evaluation.

#1. *Diflubenzuron (Dimilin):* Koc 6790; WS 0.2 ppm; a potential mutagen and carcinogen; can be excreted in milk.

#2. *Thiabendazole (MCB):* Koc 1,720; WS <50 ppm; a potential teratogen, mutagen, reproduction hazard and hazardous to wildlife.

#3. *Methomyl:* Koc 160; WS 10,000 ppm.

#4. *Carbaryl:* Koc 230; WS 40 ppm; a potential carcinogen; kills bees which are essential for pollination.

#5. *Methoxychlor:* similar to DDT in structure and effects; Koc 80,000; WS 0.003 ppm; a potential carcinogen and bird killer.

#6. *2,4-D:* Koc 20; WS 900 ppm; a potential carcinogen and teratogen.

ANSWERS

#1. *Diflubenzuron (Dimilin):* Koc 6,790; WS 0.2 ppm; a potential carcinogen and mutagen; can be excreted in milk.

Predictive Equation Analysis

$$Rf = \frac{1}{1 + (Koc)\,(\%\ OC/100)\,(ds)\,(1/\theta^{2/3} - 1)}$$

$$Rf = \frac{1}{1 + (6{,}790)\,(3/100)\,(1.5)\,(1/0.75^{2/3} - 1)}$$

$$Rf = 0.0152$$

$$Rf = \frac{1}{1 + (6{,}790)\,(1/100)\,(1.5)\,(1/0.75^{2/3} - 1)}$$

$$Rf = 0.0443$$

Fate and Transport Analysis

• The Rf values for Dimilin indicate that it should be adsorbed to soil, accumulated in soil, and not be biodegraded; should be bioaccumulated, not leached in soil, and not run off unless with soil particles. The OC values of 1 and 3 do not affect the adsorption of this chemical.

- WS of 0.2 ppm supports the Rf values and predictions.
- Koc indicates that this chemical could go either way in regards to adsorption.

Exposure Analysis

If this chemical gets in water, food-chain contamination is likely; therefore, consumption of contaminated water or food should be avoided. This chemical could be excreted in mother's milk; therefore, nursing mother's should be aware. This chemical is a potential carcinogen and mutagen and exposure to it should be avoided. This chemical is a pesticide and is applied by aircraft over residential areas—one cannot avoid inhalation and skin exposure if in the area of application. There are no data presented on volatility; however, inhalation of this chemical should be prevented, as previously discussed.

#2. *Thiabendazole (MBC):* Koc 1,720; WS <50 ppm; a potential teratogen, mutagen; reproductive hazard and hazardous to wildlife.

Predictive Equation Analysis

$$Rf = \frac{1}{1 + (Koc)\,(\%\ OC/100)\,(ds)\,(1/\theta^{2/3} - 1)}$$

$$Rf = \frac{1}{1 + (1,720)\,(3/100)\,(1.5)\,(1/0.75^{2/3} - 1)}$$

$$Rf = 0.057$$

$$Rf = \frac{1}{1 + (1,720)\,(1/100)\,(1.5)\,(1/0.75^{2/3} - 1)}$$

$$Rf = 0.1549$$

Fate and Transport Analysis

- Rf values for MBC indicate that this chemical should be adsorbed and accumulated in soil, should be bioaccumulated, and should not be leached or run off soil, unless with soil particles, and not be biodegraded. The OC values did not affect the adsorption of this chemical.

- WS indicates that this chemical could go either way in regard to adsorption, accumulation, bioaccumulation, biodegradation, leaching, and runoff.
- Koc indicates that this chemical could go either way in regard to adsorption.

Exposure Analysis

If this chemical should contaminate water or the food chain, consumption of contaminated water and food should be avoided. There are no data presented on volatility; however, inhalation of the chemical should be avoided. Contact with this chemical should be avoided because it is a potential teratogen, mutagen, can cause reproductive damage, and is hazardous to wildlife.

#3. *Methomyl:* Koc 160; Ws 10,000.

Predictive Equation Analysis

$$Rf = \frac{1}{1 + (Koc)\,(\%\ OC/100)\,(ds)\,(1/\theta^{2/3} - 1)}$$

$$Rf = \frac{1}{1 + (160)\,(3/100)\,(1.5)\,(1/0.75^{2/3} - 1)}$$

$$Rf = 0.39648$$

$$Rf = \frac{1}{1 + (160)\,(1/100)\,(1.5)\,(1/0.75^{2/3} - 1)}$$

$$Rf = 0.663398$$

Fate and Transport Analysis

- Rf of methomyl indicates that it could move in soil, but is not really mobile. The OC appears to prevent methomyl from leaching through soil. If the OC is lower than 1%, then perhaps leaching should occur. The greater the % OC, the better for adsorption in soil.
- Koc indicates that this chemical should adsorb in soil.

- WS indicates that this chemical should leach in soil.
- The data is contradictory and any prediction using the information given would be difficult. If this chemical gets into soils with OC lower than 1%, leaching should occur. Based on the WS of this chemical, it should be biodegraded.

Exposure Analysis

If methomyl gets in some soils (>1% OC), it may contaminate the aquatic environment and then the food chain. No data are presented on toxic effects, therefore, safety can not be assessed. Exposure cannot be assessed with the data given herein.

#4. *Carbaryl:* Koc 230; WS 40 ppm; a potential carcinogen; kills bees, which are needed for pollination.

Predictive Equation Analysis

$$Rf = \frac{1}{1 + (Koc)\,(\%\ OC/100)\,(ds)\,(1/\theta^{2/3} - 1)}$$

$$Rf = \frac{1}{1 + (230)\,(3/100)\,(1.5)\,(1/0.75^{2/3} - 1)}$$

$$Rf = 0.31366$$

$$Rf = \frac{1}{1 + (230)\,(1/100)\,(1.5)\,(1/0.75^{2/3} - 1)}$$

$$Rf = 0.5782$$

Fate and Transport Analysis

- Rf of carbaryl indicates that this chemical could leach in some soils and may have some mobility in others; however, leaching should not be a problem in soil with a high organic carbon content.
- WS indicates that this chemical could go either way in regard to leaching.

Exposure Analysis

If this chemical contaminated the aquatic environment and the food chain, consumption contaminated water and food should be avoided. Inhalation of this chemical should be prevented because it is a carcinogen. If bees are important for pollination, avoid the use of carbaryl, because bee kill may result.

#5. *Methoxychlor:* Koc 80,000; chemical structure similar to DDT; WS 0.003 ppm; a potential carcinogen and hazardous to wildlife.

Predictive Equation Analysis

$$Rf = \frac{1}{1 + (Koc)\,(\%\ OC/100)\,(ds)\,(1/\theta^{2/3} - 1)}$$

$$Rf = \frac{1}{1 + (80,000)\,(3/100)\,(1.5)\,(1/0.75^{2/3} - 1)}$$

$$Rf = 0.00131216$$

$$Rf = \frac{1}{1 + (80,000)\,(1/100)\,(1.5)\,(1/0.75^{2/3} - 1)}$$

$$Rf = 0.003926$$

Fate and Transport Analysis

- The Rf values of methoxychlor indicate that this chemical should not leach and run off unless with soil particles, and should not be biodegraded, should be adsorbed and accumulated in soil, and be bioaccumulated.
- The Koc value of 80,000 supports the above Rf prediction.
- WS of 0.003 ppm also supports the above prediction.

Exposure Analysis

If this chemical is released in the environment it could result in the contamination of water and the food chain. Consumption of contaminated water or food should be avoided because this chemical is like DDT and could be carcinogenic. This chemical could be hazardous if inhaled; however, no data are presented to indicate volatility. If similar to DDT, volatility could be a problem. This chemical, like DDT, is hazardous to birds.

#6. *2,4-D:* Koc 20; WS 900 ppm; a potential carcinogen and teratogen.

Predictive Equation Analysis

$$Rf = \frac{1}{1 + (Koc)\,(\%\ OC/100)\,(ds)\,(1/\theta^{2/3} - 1)}$$

$$Rf = \frac{1}{1 + (20)\,(3/100)\,(1.5)\,(1/0.75^{2/3} - 1)}$$

$$Rf = 0.84$$

$$Rf = \frac{1}{1 + (20)\,(1/100)\,(1.5)\,(1/0.75^{2/3} - 1)}$$

$$Rf = 0.959$$

Fate and Transport Analysis

- The Rf of 2,4-D indicates that this chemical could leach in soil, run off with soil and water, should be biodegraded, should not be adsorbed or accumulated in soil, and should not be bioaccumulated.
- Koc indicates that adsorption should not occur.
- WS indicates that this chemical could leach and run off; therefore, contamination to the aquatic environment is likely if the chemical is not biodegraded rapidly.

Exposure Analysis

If the aquatic environment is contaminated, then consumption of contaminated water and food should be avoided. This chemical is a suspected carcinogen and teratogen and consumption of contaminated water and food should be avoided. Although no data are given on volatility, inhalation of 2,4-D residues should be avoided. As mentioned earlier, 2,4-D residues could show up in mother's milk; therefore, nursing mothers should take precautions against inhalation.

REFERENCES

[1] Helling, Charles S. and Benjamin C. Turner, "Pesticide Mobility: Determination by Soil Thin-Layer Chromatography", *Science*, Vol. 162, pp.562-563, 1968.
[2] Hamaker, John W., "Interpretation of Soil Leaching Experiments, Dynamics of Pesticides in the Environment", Plenum Publishing Corporation, pp. 21-29, 1979.

Chapter 9
Octanol-Water Partition Coefficient

The octanol-water partition coefficient (Kow) is the term used to show the ratio of how an organic chemical partitions between an octanol phase and a water phase when at equilibrium. It is also expressed as the distribution of an organic chemical between two polar solvents—water and octanol. Octanol is the solvent of choice, although other organic solvents may be used. The determination for Kow is expressed by the following equation.

$$Kow = \frac{\text{concentration of the chemical in octanol}}{\text{concentration of the chemical in water}} \quad \text{eq. 9-1}$$

Kow is an indicator of a chemical's ability to partition in fat tissue: that is, to bioaccumulate in animals. The higher the value for Kow, the greater the affinity to be lipo-soluble. Kow also serves as an indicator of hydrophobicity; that is, the affinity to be soluble in water. The lower the Kow, the less lipo-soluble and the greater the water solubility.

When Kow data does not exist for chemicals, there are many ways to determine a chemical's Kow. Two such ways have been discussed—WS and Koc.

Another correlation method is the measure of retention time of a chemical when a high-pressure liquid chromatograph (HPLC) is used. The time it takes for the chemical to be passed through an HPLC system from injection to detection has been found to be linearly related (the logarithm of the retention time). The HPLC system should have a column with an octanol stationary phase and a micellar mobile phase. The Kow could be calculated if the retention time (the time it takes from injection to detection) was known from the HPLC system. Column preparation for the HPLC system may cause errors in the correlation, as well as chemical purity. Correlations have been reported by Vieth [1], Janini [2], and Chiou [3].

Although we do not have a HPLC, we can still make predictions. Kow can be predicted from WS, Koc, and BCFs data. Now let us say we have a Kow and no other data— just what can we do with it? We can use mathematical equations— sometimes called mathematical models— to predict WS, Koc, and BCFs. The following equations can be used to predict WS, Koc, and BCFs.

$$\log WS = 4.184 - 0.922 \log Kow \quad [4] \qquad \text{eq. 9-2}$$

$$\log Koc = 1.377 + 0.544 \log Kow \quad [4] \qquad \text{eq. 9-3}$$

$$\log BCF_{(f)} = -1.495 + 0.935 \log Kow \quad [5] \qquad \text{eq. 9-4}$$

$$\log BCF_{(t)} = -0.973 + 0.767 \log Kow \quad [5] \qquad \text{eq. 9-5}$$

The numerical values in these equations are standard. Let us use these equations and find WS, Koc, and BCFs.

Example: 2,4-D has a Kow of 37 = log Kow 1.568 [1]: Solve for WS; Koc; BCFs; a potential carcinogen.

$$\log WS = 4.184 - 0.922 \log Kow$$
$$= 4.184 - 0.922 \times 1.57$$
$$= 2.738$$
$$WS = 547.02 \text{ ppm}$$

$$\log Koc = 1.377 + 0.544 \log Kow$$
$$= 1.377 + 0.544 \times 1.57$$
$$= 2.2299$$
$$Koc = 169.82$$

$$\log BCF_{(f)} = -1.495 + 0.935 \log Kow$$
$$= -1.495 + 0.935 \times 1.57$$
$$= -0.029$$
$$BCF_{(f)} = 0.936$$

$$\log BCF_{(t)} = -0.973 + 0.767 \log Kow$$
$$= -0.973 + 0.767 \times 1.57$$
$$= 0.2297$$
$$BCF_{(t)} = 1.697$$

Now we have obtained the following values for 2,4-D: WS 547.016 ppm; Koc 3.342; Kow 37; $BCF_{(f)}$ 0.936; $BCF_{(t)}$ 1.697. Using the predictive method described in Part I and given under solubility, 2,4-D would be assessed as follows:

- WS of 2,4-D indicates that leaching, runoff, and biodegradation may occur, and adsorption in soil, accumulation in soil, and bioaccumulation should not occur.
- Koc indicates that the chemical should not adsorb to soil.
- Kow indicates that bioaccumulation should not occur.
- BCFs indicate that this chemical should not bioaccumulate; however, residues could be expected in the food chain.
- Exposure would be by consumption of contaminated food and water, if such occurred. Volatility could be a route of exposure; however, there are no data presented on volatility; therefore, this route of exposure cannot be considered. 2,4-D is a potential carcinogen; therefore, no exposure is acceptable. The aggregate, antagonistic, and synergistic effects with other hazardous chemicals cannot be assessed.
- Although the values calculated using Kow are different from those calculated by using WS and Koc, the end analysis is the same. Remember that there is a degree of error using these equations, but the end analysis is acceptable.

EXERCISE

Solve each of the following chemicals for WS, Koc, and BCFs; write a fate and transport prediction analysis for each; and write an exposure analysis for each. Use Table 10-1, Chapter 10, under solubility to help make predictions.

#1. *Carbaryl:* Kow 230 [1]; a potential carcinogen; kills bees.
#2. *Captan:* Kow 224 [1]; a potential mutagen, teratogen, neurotoxin, and fetotoxin; hazardous to wildlife. This chemical is similar to and may break down to thalidomide which causes fetal abnormalities.
#3. *DBCP, also 1,2-Dibromo-3-chloropropane:* Kow 129; volatile [1]; a potential carcinogen; may cause sterility.
#4. *9-Methylanthracene:* Kow 117,000 [1].
#5. *Pentachlorophenol:* Kow 900 [1]; a potential teratogen and carcinogen.

ANSWERS

#1. Carbaryl: Kow 230 = log Kow 2.35; may cause cancer.

Predictive Equation Analysis

$$\log WS = 4.184 - 0.922 \log Kow$$
$$= 4.184 - 0.922 \times 2.36$$
$$= 2.017$$
$$WS = 104.06 \text{ ppm}$$

$$\log Koc = 1.377 + 0.544 \log Kow$$
$$= 1.377 + 0.544 \times 2.36$$
$$= 2.6608$$
$$Koc = 457.97$$

$$\log BCF_{(f)} = -1.495 + 0.935 \log Kow$$
$$= -1.495 + 0.935 \times 2.36$$
$$= 0.7116$$
$$BCF_{(f)} = 5.147$$

$$\log BCF_{(t)} = -0.973 + 0.767 \log Kow$$
$$= -0.973 + 0.767 \times 2.36$$
$$= 0.837$$
$$BCF_{(t)} = 6.97$$

Fate and Transport Analysis

- WS of carbaryl indicates that it could go either way regarding leaching, runoff, adsorption to soil, accumulation in soil, biodegradation, and bioaccumulation.
- Koc indicates that adsorption in soil should not occur.
- Kow indicates that bioaccumulation should not occur.
- BCFs indicate that residues should be found in the food chain.
- Koc and Kow supports solubility, therefore leaching, runoff, and biodegradation could occur. Bioaccumulation should not occur. Since this chemical can be mobile in the environment, food-chain contamination could occur.

Exposure Analysis

This chemical is a potential carcinogen; therefore, consumption of contaminated food and water should be avoided. Inhalation of volatile residues could be a problem; however, no data have been given on volatility to assess inhalation. This chemical can kill bees; therefore, important pollination may be prevented. This chemical can be hazardous to humans and the environment.

#2. *Captan:* Kow 224 = log Kow 2.35; a potential teratogen, fetotoxin, mutagen, and neurotoxin; hazardous to wildlife. The chemical structure of this chemical is similar to thalidomide, which has caused birth defects. This chemical may break down to thalidomide.

Predictive Equation Analysis

$$\log WS = 4.184 - 0.922 \log Kow$$
$$= 4.184 - 0.922 \times 2.35$$
$$= 2.017$$
$$WS = 104.06 \text{ ppm}$$

$$\log Koc = 1.377 + 0.544 \log Kow$$
$$= 1.377 + 0.544 \times 2.35$$
$$= 2.6554$$
$$Koc = 452.27$$

$$\log BCF_{(f)} = -1.495 + 0.935 \log Kow$$
$$= -1.495 + 0.935 \times 2.35$$
$$= 0.702$$
$$BCF_{(f)} = 5.04$$

$$\log BCF_{(t)} = -0.973 + 0.767 \log Kow$$
$$= -0.973 + 0.767 \times 2.35$$
$$= 0.82$$
$$BCF_{(t)} = 6.75$$

Fate and Transport Analysis

- WS of captan indicates that it could go either way regarding leaching, runoff, adsorption in soil, accumulation in soil, biodegradation, and bioaccumulation.
- Koc indicates that this chemical should not adsorb in soil.
- Kow indicates that this chemical should not bioaccumulate.
- BCFs indicate that bioaccumulation should not occur; however, residues could be found in the food chain.

- Koc and Kow indicate solubility; therefore leaching, runoff, and biodegradation could occur.
- Bioaccumulation should not occur. Because of all the adverse effects that can be caused by this chemical, it should be avoided.

Exposure Analysis

Because of the adverse effects that can be caused by captan, consumption of contaminated food and water should be avoided. Inhalation of volatile residues should be prevented if this chemical is volatile; however, no data have been presented on volatility. This chemical and one of its break down products are hazardous to humans and the environment.

#3. *1,2-Dibromo-3-chloropropane also DBCP:* Kow 129 = log Kow 2.11; volatile; a potential carcinogen; may cause sterility.

Prediction Equation Analysis

$$\log WS = 4.184 - 0.922 \log Kow$$
$$= 4.184 - 0.922 \times 2.11$$
$$= 2.239$$
$$WS = 173 \text{ ppm}$$

$$\log Koc = 1.377 + 0.544 \log Kow$$
$$= 1.377 + 0.544 \times 2.11$$
$$= 2.525$$
$$Koc = 334.84$$

$$\log BCF_{(f)} = -1.495 + 0.935 \log Kow$$
$$= -1.495 + 0.935 \times 2.11$$
$$= 0.478$$
$$BCF_{(f)} = 3.01$$

$$\log BCF_{(t)} = -0.973 + 0.767 \log Kow$$
$$= -0.973 + 0.767 \times 2.11$$
$$= 0.645$$
$$BCF_{(t)} = 4.419$$

Fate and Transport Analysis

- WS of DBCP indicates that this chemical can go either way regarding leaching, runoff, adsorption in soil, accumulation in soil, biodegradation, and bioaccumulation.

- Koc indicates that this chemical should not be adsorbed in soil.
- Kow indicates that the chemical should not bioaccumulate.
- BCFs indicate that bioaccumulation should not occur; however, residues may be found in the food chain.
- This chemical is volatile. Inhalation of this carcinogenic and sterility chemical must be prevented. Volatile residues are subject to phototransformation and fallout onto waters and foods.

Exposure Analysis

This chemical can contaminate food and water; therefore, consumption of its residues should be prevented. Inhalation— which is the most probable route of human exposure— must be prevented. Adverse effects can be expected from this chemical.

#4. *9-Methylanthracene:* Kow 117,000 = log Kow 5.068.

Predictive Equations Analysis

$$\log WS = 4.184 - 0.922 \log Kow\ 5.068$$
$$= 4.184 - 0.922 \times 5.068$$
$$= -0.4886$$
$$WS = 0.325\ ppm$$

$$\log Koc = 1.377 + 0.544 \log Kow$$
$$= 1.377 + 0.544 \times 5.068$$
$$= 4.1339$$
$$Koc = 13{,}614.196$$

$$\log BCF_{(f)} = -1.495 + 0.935 \log Kow$$
$$= -1.495 + 0.935 \times 5.068$$
$$= 3.2436$$
$$BCF_{(f)} = 1{,}752.18$$

$$\log BCF_{(t)} = -0.973 + 0.767 \log Kow$$
$$= -0.973 + 0.767 \times 5.068$$
$$= 2.914$$
$$BCF_{(t)} = 820.646$$

Fate and Transport Analysis

- WS of 9-Methylanthracene indicates that this chemical should be adsorbed in soil, accumulated in soil, and bioaccumulated, and should not leach, runoff, or be biodegraded.

- Koc indicates that chemical should adsorb and accumulate in soil.
- Kow indicates that this chemical should bioaccumulate in the food chain.
- BCFs indicate that the chemical should bioaccumulate, and residues should be expected in the food chain.

Exposure Analysis

This chemical should be persistent in any environmental compartments (air, water, soil, plants and animals). Exposure would be by consumption of contaminated food and water. Inhalation of volatile residues could be a problem; however, there are no data presented on volatility to assess inhalation.

#5. *Pentachlorophenol:* Kow 900 = log Kow 2.95; a potential teratogen and carcinogen.

Predictive Equation Analysis

$$\log WS = 4.184 - 0.922 \log Kow$$
$$= 4.184 - 0.922 \times 2.95$$
$$= 1.464$$
$$WS = 28.8 \text{ ppm}$$

$$\log Koc = 1.377 + 0.544 \log Kow$$
$$= 1.377 + 0.544 \times 2.95$$
$$= 2.9818$$
$$Koc = 958.959$$

$$\log BCF_{(f)} = -1.495 + 0.935 \log Kow$$
$$= -1.495 + 0.935 \times 2.95$$
$$= 1.26$$
$$BCF_{(f)} = 18$$

$$\log BCF_{(t)} = -0.973 + 0.767 \log Kow$$
$$= -0.973 + 0.767 \times 2.95$$
$$= 1.29$$
$$BCF_{(t)} = 19.48$$

Fate and Transport Analysis

- WS of pentachlorophenol indicates that this chemical could go either way regarding leaching, runoff, adsorption in soil, accumulation in soil, biodegradation, and bioaccumulation.
- Koc indicates that this chemical should not adsorb in soil.
- Kow indicates that chemical can go either way regarding bioaccumulation.
- BCFs indicate that the chemical should not bioaccumulate; however, residues could be found in the food chain based on all of the above data.

Exposure Analysis

Exposure would be by consumption of contaminated food and water. Inhalation of volatile residues should be prevented if the chemical is found to be volatile; however, no data have been presented on volatility to assess this exposure. The chemical is a potential teratogen and carcinogen, therefore, any exposure should be prevented.

REFERENCES

[1] Vieth, G. D., Austin, N. M. and Morris, R. T., "A Rapid Method for Estimating Log P for Organic Chemicals," *Water Res.*, 13, pp 43-47, 1979.
[2] Janini, G. M. and S. A. Attari, "Determination of Partition Coefficient of Polar Solutes in Octanol/Micellar Solutions," Department of Chemistry, University of Kuwait, Kuwait, *Analytical Chemistry*, pp 659-661, 1983.
[3] Chiou, Cary T. and David W. Schmedding, "Measurement and Interrelation of Octanol-Water Partition Coefficient and Water Solubility of Organic chemicals, Test Protocol for Environmental Fate and Movement of Toxicants", *Proceeding of a Symposium, Association of Official Analytical Chemist*, 94th Annual Meeting, pp 28-42, September 21,22, 1980.
[4] Kenaga, E. E. and C. A. I. Goring, "Relationship Between Water Solubility, Soil Sorption, Octanol-Water Partitioning, and Concentration of Chemicals in Biota", *Special Technical Publication 707, American Society for Testing and Materials*, 1980.
[5] Layman, Warren J., William F. Reehl and David H. Rosenblatt, Ph.D., *Handbook of Chemical Property Estimation Methods*, McGraw-Hill Book Company, 1982.

Chapter 10
Water Solubility

Water solubility (WS) data is probably the most important physical and chemical property of a chemical, because it is the most abundant type of data. Also what makes it important is that WS can be used to predict biodegradation, metabolism, runoff, accumulation in soil, bioaccumulation, octanol water partition coefficient (Kow), adsorption coefficient (Koc), bioaccumulation factor (BCF), and retardation factor/retention factor (Rf) (leaching).

If WS data on a chemical does not exist, then you can use chemical structures for hydrocarbons and halo hydrocarbons to predict WS. If data exists on Kow, Koc, or BCFs, then you can use regression equations/ mathematical equations to predict WS; vice versa is also true. Why do we want to make predictions? If only limited data are available on a chemical and you want to know what happens to a chemical in the environment, or you want to know your potential for exposure to a chemical, then you can predict such.

WS is defined as the amount of chemical that will dissolve in water (pure water) at ambient temperature or some pre selected temperature. WS can be expressed in units of weight (ppm, ppb, g/kg, etc.) or in units of volume (mg/L, ug/L, etc.).

It is easy to use WS to predict other environmental parameters, such as Koc, Kow, Rf, $BCF_{(f)}$ and $BCF_{(t)}$. It is far more difficult to determine WS using the chemical structure technique. This method is for the advanced scientists. However, I will briefly discuss the technique and try to present it in a simplified manner. I will give an example so that you will be aware of the existence of the chemical structure technique. For full details you can refer to *Handbook of Chemical Property Estimates Methods [1]*. Irmann [2] developed the mechanism for predicting water solubility of a chemical using the characteristics of its chemical structure. These techniques only apply to hydrocarbons and halo hydrocarbons, thus are limited for our use. The method differs for liquids, solids, and gases. Equations and discussions are presented below.

The basic equation for WS of a liquid at 25°C is presented below.

$$-\log S = X + Y_i N_i + Z_j N_j \qquad \text{eq. WS 10-1}$$

X, Y, and Z are obtainable from tables in the handbook. (1) X is the type chemical compound, Y is the location of specific atoms (i.e., C, H, F, Cl, Br and I), and Z is for the structural elements. The equation is seen and explained below.

Type compound --------------------= X
Number of specific atoms --------= $Y_i N_i$
Structural elements ----------------= $Z_j N_j$
add these to get----------------------= -log S,
convert log S to get the whole number for WS.

Example: let $X = 1$, $Y_i N_i = 1/2 + 1/2 = 1$, and $Z_j N_j = 1$. Substitute these into the equation and you have

-log S = 1+1+1
-log S = 3
WS   = $1 \times 10^{-3} \times 1,000,000$ to get mg/L,
therefore, the whole number is WS = 1,000 mg/L.

The equation for a solid at 25°C is presented below.

$$-\log S_{sol} = -\log S + 0.0095 \, (TM-25) \qquad \text{eq. WS 10-2}$$

The -log S is the right side of equation WS-1, TM is the melting point of the solid, 25 represents 25°C, and the number 0.0095 is a standard factor. Consider the TM to be 100 and the -log S = 3 as above and we have the following:

$-\log S_{sol} = 3 + 0.0095\ (100 - 25)$

$\qquad\quad = 3.7125$

WS $\quad = 1.9386 \times 10^{-4} \times 1{,}000{,}000$ to get ug/L,

therefore, the whole number is WS $= 193.86526$ ug/L.

The equation of a gas is the same as for a liquid, if the chemical is a gas at 1 atmosphere (atm) (normal vapor pressure) and at 25°C. In other words, the solubility of a liquified gas is at the vapor phase of the two phases (chemical(gas) and water).

Remember, these were only presented for those who would like to try the chemical structure technique. For simplicity, from now on I will refer to the regression equations as mathematical equations or equations. Sometimes they have been referred to as mathematical models. These equations will use WS to predict Koc, Kow, and BCFs and are presented below.

$\log Koc \quad = 3.64 - 0.55 \log WS$ [1] eq. WS 10-3

$\log Kow \quad = 4.158 - 0.8 \log WS$ [3] eq. WS 10-4

$\log BCF_{(f)} = 2.791 - 0.564 \log WS$ [1] eq. WS 10-5

$\log BCF_{(t)} = 2.183 - 0.629 \log WS$ [1] eq. WS 10-6

The numerical values in these equations are standard factors. Recall in Part I that the (f) and (t) after BCF are for a continuous flow and static fish study, respectively.

Now use these equations to find the unknowns Koc, Kow, and BCFs. 2,4-Dichlorophenoxyacetic acid (2,4-D) has a WS of 900 ppm as given in Part I, which is equivalent to log WS of 2.954. Solve for Koc, Kow, and BCFs.

$\log Koc \quad = 3.64 - 0.55 \log WS$

$\qquad\qquad = 3.64 - 0.55\ (2.954)$

$\qquad\qquad = 2.0153$

Koc $\qquad = 103.55$

$\log Kow \quad = 4.158 - 0.80 \log WS$

$\qquad\qquad = 4.158 - 0.80\ (2.954)$

$\qquad\qquad = 1.7948$

Kow $\qquad = 62.34$

$$\log \text{BCF}_{(f)} = 2.791 - 0.564 \log \text{WS}$$
$$= 2.791 - 0.564 \ (2.954)$$
$$= 1.1249$$
$$\text{BCF}_{(f)} = 13.33$$

$$\log \text{BCF}_{(t)} = 2.183 - 0.629 \log \text{WS}$$
$$= 2.183 - 0.629 \ (2.954)$$
$$= 0.3249$$
$$\text{BCF}_{(t)} = 2.113$$

We now have the following values for 2,4-D: WS = 900 ppm; Koc = 103.58; Kow = 62.34; $\text{BCF}_{(f)}$ = 13.33; $\text{BCF}_{(t)}$ = 2.113. Using the predictive method described in Part I, 2,4-D would be assessed as follows:

- WS and Koc indicates that leaching, runoff, and biodegradation could occur, and that adsorption in soil and accumulation in soil should not occur.
- Kow and BCFs indicates that bioaccumulation in the food chain should not occur; however, residues could be expected in the food chain based on values for BCFs.
- Exposure would be by consumption of contaminated food and water. Volatility could be another route of exposure; however, there are no data presented on volatility, therefore, it cannot assess if inhalation is another route exposure. The aggregate, antagonistic, and synergistic effects are unknown; therefore, effects due to exposure with other hazardous chemicals cannot be assessed. In a study by Yip and Ney [4], cows grazing on grass treated with 2,4-D, were detected to have 2,4-D as a protein-bound residue in milk. 2,4-D is a known carcinogen; therefore, if nursing mother's have been exposed to this chemical, they should have their milk analyzed for the presence or absence of 2,4-D. Esters of 2,4-D have been detected in air from samples taken in residential homes [5], thus nursing mothers should be made aware. Volatile residues of 2,4-D can be inhaled, and this is likely to result in residues of 2,4-D in mother's milk and human tissue.

EXERCISE

Solve each of the following for Kow, Koc, and BCFs; write a fate and transport prediction analysis for each; and write an exposure analysis for each. The following tables taken from the book will also help.

Table 10-1. Predicting Where the Chemical Goes.

PREDICTIVE TECHNIQUES		RANGES	
WS-ppm	<10	10-1,000	>1,000
Hydrolysis, $T_{1/2}$ days	>90	30-90	<30
Photolysis, $T_{1/2}$ days	>90	30-90	<30
VP-mm Hg	<0.000001	0.000001-0.01	>0.01
Koc	>10,000	1,000-10,000	<1,000
Rf	<0.34	0.34-0.75	>0.75
Kow	>1,000	500-1,000	<500
Soil, $T_{1/2}$ months	6	2-6	<2
FATE AND TRANSPORT			
Soluble	n*	either way	yes
Hydrolyses	n	either way	yes
Photolyses	n*	either way	yes
Volatile	n	either way	yes
Adsorbs	yes	either way	n
Leaches	n	either way	yes
Runoff	n	either way	yes
Bioaccumulation	yes	either way	n
Persists	yes	either way	n
Biodegraded	slowly	either way	yes
Is metabolized	slowly	either way	yes

* n denotes negligible

#1. *Dichlorvos (DDVP or Vapona):* WS 10,000 ppm; volatile [4]; a potential oncogen.

#2. *Diazinon:* WS 40 ppm [4]; a potential avian hazard (death in wildlife).

#3. *Benomyl:* a potential mutagen and teratogen; causes reproductive hazards and is hazardous to wildlife; degrades to many chemicals, one of which is methyl 2-benzimidazolecarbamate (MBC), a potential carcinogen.

#4. *Maneb:* moderately soluble in water; moderately toxic by ingestion and inhalation; a potential oncogen and teratogen; breaks down to ethylenethiourea (ETU), a potential carcinogen.

#5. *Pronamide:* WS-15 ppm [3]; an oncogen.

ANSWERS

#1. *Dichlorvos;* WS = 10,000 ppm = log 4, volatile, and an oncogen.

Predictive Equation Analysis

$$\log K_{oc} = 3.64 - 0.55 \log WS$$
$$\log K_{oc} = 3.64 - 0.55 \times 4$$
$$\log K_{oc} = 1.44$$
$$K_{oc} = 27.5$$

$$\log K_{ow} = 4.158 - 0.8 \log WS$$
$$\log K_{ow} = 4.158 - 0.8 \times 4$$
$$\log K_{ow} = 0.958$$
$$K_{ow} = 9.08$$

$$\log BCF_{(f)} = 2.791 - 0.564 \log WS$$
$$\log BCF_{(f)} = 2.791 - 0.564 \times 4$$
$$\log BCF_{(f)} = 0.535$$
$$BCF_{(f)} = 3.43$$

$$\log BCF_{(t)} = 2.183 - 0.629 \log WS$$
$$\log BCF_{(t)} = 2.183 - 0.629 \times 4$$
$$\log BCF_{(t)} = -0.333$$
$$BCF_{(t)} = 0.456$$

Fate and Transport Analysis

- WS indicates that this chemical could leach, runoff, and be biodegraded, should not be adsorbed or accumulated in soil, and not be bioaccumulated in animals.
- Koc indicates that this chemical should not adsorb or accumulate in soil.
- Kow indicates that this chemical should not be bioaccumulated in animals; this is supported by BCF values.
- BCFs indicate that this chemical does not bioaccumulate, probably because it is metabolized in animals. Residues of parent chemical and possible metabolites may be found in tissue. This type chemical (an organo phosphate) is usually toxic, therefore, death in animals may occur.
- Volatility indicates that this chemical could be inhaled and the volatile residues could fallout and contaminate the food chain. Due to the lack of data, the formation of phototransformation products cannot be assessed.

Exposure Analysis

Dichlorvos is a potential oncogen; therefore, any exposure should be eliminated. Exposure would be by inhalation of the volatile residues, by residues on the skin from fallout of the volatile residues, and by consumption of contaminated food and water. This chemical is an organo phosphate, and organo phosphates are toxic, thus food chain contamination and human exposure to this chemical should be prevented.

#2. *Diazinon:* WS 40 ppm = log WS 1.6; a bird killer.

Predictive Equation Analysis

$$\log K_{oc} = 3.64 - 0.55 \log WS$$
$$\log K_{oc} = 3.64 - 0.55 \times 1.6$$
$$\log K_{oc} = 2.76$$
$$K_{oc} = 575.44$$

$$\log K_{ow} = 4.158 - 0.8 \log WS$$
$$\log K_{ow} = 4.158 - 0.8 \times 1.6$$
$$\log K_{ow} = 2.878$$
$$K_{ow} = 755.09$$

$$\log BCF_{(f)} = 2.791 - 0.564 \log WS$$
$$\log BCF_{(f)} = 2.791 - 0.564 \times 1.6$$
$$\log BCF_{(f)} = 1.89$$
$$BCF_{(f)} = 77.37$$

$$\log BCF_{(t)} = 2.183 - 0.629 \log WS$$
$$\log BCF_{(t)} = 2.183 - 0.629 \times 1.6$$
$$\log BCF_{(t)} = 1.18$$
$$BCF_{(t)} = 15.02$$

Fate and Transport Analysis

- WS indicates that this chemical could go either way in regards to leaching, runoff, adsorption, accumulation, biodegradation, and bioaccumulation.
- Koc and Kow indicates the same as WS.

- BCFs indicate that bioaccumulation should not occur; however, residues could be expected in the food chain. Metabolism in animals may have occurred as indicated by the low values for the BCFs. This chemical is an organo phosphate and should be considered toxic; in fact, it has killed birds.

Exposure Analysis

Residues of diazinon may end up in water, plants, and animals. Consumption of contaminated food and water should be prevented. Data on volatility are not given, but inhalation exposure to possible volatile residues should be prevented. Wildlife—especially avian exposure—must be prevented or death may occur.

#3. *Benomyl:* a potential mutagen and teratogen; causes hazards to reproductive system and wildlife; breaks down to many chemicals, one is methyl 2-benzimidazolecarbamate (MBC) (a potential carcinogen).

Predictive Equation Analysis

No predictions here, because no data are given.

Fate and Transport Analysis

- Benomyl breaks down to many chemicals; therefore, the prediction is that it is broken down in the environment and in plants and animals. Benomyl's break-down product MBC is also hazardous in the environment and to humans. Because benomyl is broken down, the prediction is that: it is water soluble and will be mobile in the environment; it should not bioaccumulate in animals; and it should not adsorb or accumulate in soil. Benomyl and its break-down products could contaminate water and the food chain.

Exposure Analysis

Benomyl is a potential mutagen and teratogen, a reproductive hazard, and a hazard to wildlife. Benomyl also breaks down to MBC a potential carcinogen; therefore, any exposure is a hazard. Consumption of contaminated water and food, and inhalation of any benomyl residues (no data presented on volatility) is hazardous to humans and the environment.

#4. *Maneb:* WS-moderately soluble in water; a potential oncogen and teratogen; moderately toxic by ingestion and inhalation; it breaks down to ethylenethiourea (ETU) (a potential carcinogen).

Predictive Equation Analysis

No predictions here, because no data are given.

Fate and Transport Analysis

• Maneb is moderately soluble; therefore, the prediction is that it should leach, runoff, and be biodegraded; it should not be adsorbed or accumulated in soil; and it should not be bioaccumulated in animals. Since maneb breaks down (for example, to ETU), the prediction is that it is broken down in the environment, and in plants and animals.

Exposure Analysis

Maneb could contaminate water and the food chain, therefore, consumption could be a problem. This chemical is a known oncogen and teratogen, therefore, no exposure should be acceptable. The breakdown product, ETU, is a potential carcinogen and exposure to it should be prevented. If the parent chemical contaminates water or the food chain, then so can ETU. Data on volatility are not presented; however, because of the toxic properties of maneb and ETU, any exposure should be prevented. This chemical could be hazardous to humans and wildlife.

#5. *Pronamide:* WS 15 = log WS 1.176; a potential oncogen.

Predictive Equation Analysis

$$\log Koc = 3.64 - 0.55 \log WS$$
$$\log Koc = 3.64 - 0.55 \times 1.176$$
$$\log Koc = 2.99$$
$$Koc = 984.35$$

$$\log Kow = 4.158 - 0.8 \log WS$$
$$\log Kow = 4.158 - 0.8 \log 1.176$$
$$\log Kow = 3.22$$
$$Kow = 1648.92$$

$\log BCF_{(f)} = 2.791 - 0.564 \log WS$
$\log BCF_{(f)} = 2.791 - 0.564 \times 1.176$
$\log BCF_{(f)} = 2.13$
$\quad BCF_{(f)} = 134$

$\log BCF_{(t)} = 2.183 - 0.629 \log WS$
$\log BCF_{(t)} = 2.183 - 0.629 \times 1.176$
$\log BCF_{(t)} = 1.443$
$\quad BCF_{(t)} = 27$

Fate and Transport Analysis

- WS indicates that this chemical could go either way in regards to leaching, runoff, biodegradation, adsorption or accumulation in soil, and bioaccumulation.
- Koc indicates that this chemical could be adsorbed in soil.
- Kow indicates that this chemical could be bioaccumulated.
- BCFs indicate that bioaccumulation should not occur; however, residues could be expected in the food chain.
- The data are contradictory; however, exposure to this chemical should be avoided because it is a potential oncogen.

Exposure Analysis

#5. Pronamide residues could contaminate water and the food chain; therefore, because this chemical is a potential oncogen, exposure to any residue should be prevented. Data on volatility are not given; however, this chemical is a potential oncogen, and inhalation should be prevented. This chemical could cause food-chain contamination; therefore, exposure to humans and wildlife should be prevented.

REFERENCES

[1] Layman, Warren J., William F. Reehl and David H. Rosenblatt, Ph.D., *Handbook of Chemical Property Estimation Methods*, McGraw-Hill Book Company, 1982.
[2] Irmann, F., "A Simple Correlation Between Water Solubility and Structure of Hydrocarbons and Halohydrocarbons," *Chem. Ing. Tech., 37, 789-98, 1965.*
[3] Kenaga, E. E. and C. A. I. Goring, "Relationship Between Water Solubility, Soil Sorption, Octanol-Water Partitioning, and Concentration of Chemicals in Biota", *Special Technical Publication 707, American Society for Testing and Materials,* 1980.

[4] Yip, George and Ronald E. Ney, Jr., "Analysis of 2,4-D Residues in Milk and Forage," *Weeds, Journal of the Weed Society of America*, vol. 14, pp.167-170, April 1966.
[5] Environmental Protection Agency, *Nonoccupational Pesticide Exposure Study (NOPES)*, EPA/600/3-90/003, January 1990.

Chapter 11
Sorption In Soil

The principles of chemical sorption to soils and sediments have been studied for years. Soils are complex, with different types of interactions that could take place as a result of varying percentages of organic matter, organic carbon, clays, microbes, etc. Chemical sorption to soil can occur by ion exchange, magnetic interactions, microbial, hydrophobic bonding, electron transfer, etc. Other sources outside of the soils that affect chemical sorption in soil are the amount and rate of precipitation, temperature, vapor pressure of the chemical, etc. Considering all the reactive possibilities, it would difficult to—if not impossible without the aid of computer modeling—to predict chemical fate and transport in soil. Therefore, I have chosen the one mechanism that is most prevalent, that is, sorption in soils by organic carbon (OC) or organic matter(OM). Even when using the predictive techniques that I will discuss, different percentages of OM or OC cannot be considered.

Chemical sorption to soil occurs when a chemical in solution is partitioned from the solvent (solute) into the sorbent (in our case soil). Most organic compounds with low water solubility are hydrophobic or liposoluble and can be adsorbed to soil's OC or OM. Adsorption coefficients Koc or Kom are not the same. Koc is adsorption to soils organic carbon, while Kom is adsorption to soil's organic matter. They are close, but different and not interchangeable. The ratio of Koc to Kom in soil varies; however, a standard value of 1.724 is usually used to correct for any variances. To convert Koc to Kom, or vice versa, is expressed by the following equation.

$$Koc = 1.724 \times Kom \hspace{4cm} \text{eq. 11-1}$$

The adsorption of aromatic organic chemicals (benzene ring chemicals) in soils or in sediment by OM or OC has been referred to as van der Walls forces.

Our discussions will be limited to Koc. Sorption to OC increases as chemical solubility in water decreases. As a chemical becomes less polar or the number of carbon atoms increases, the chemical tends to become less water soluble and adsorbs to soil. Sorption to OC is expressed by the following equation.

$$Koc = Kd \,/\, \% \, OC \times 100 \qquad\qquad \text{eq. 11-2}$$

Kd is considered a constant and is considered to be 1, if no values have been given. This equation is presented to enable the reader to visualize that as the percentage of OC changes, so does the adsorption coefficient (Koc) to soil. If the soil is sand and contains no (zero) OC then Koc is 0, thus leaching of almost any chemical is inevitable.

There are many exceptions where chemicals do not adsorb to OC or OM; for example, paraquat (a pesticide) adsorbs to soil's cation exchange capacity (CEC). Other exceptions include, but are not limited to, adsorption to clays with large surface areas and polymerization with soils colloid particles. These are considered when using the equations that will be presented. Soil adsorption (Koc) may be determined in the laboratory by using a shake flask technique. This technique uses water, soil, and the chemical to be studied. The chemical is distributed between either the soil's OC and the water phase, or just in OC or water. This relationship is expressed by the following equation.

$$Koc = \frac{\text{ug chemical adsorbed/g soil OC}}{\text{ug chemical soluble/g water}} \qquad \text{eq. 11-3}$$

Water does not have to be the solvent; any solvent can be used. For our purpose water is the solvent of choice, because it occurs naturally as rain, snow, etc., and is considered the universal solvent. Solvents other than water will be used for liquid spills, etc.

Koc can be calculated and predicted from WS as previously discussed. Now let us consider that we only know the Koc and have no other data. The following equations can be used to predict WS, Kow, and BCFs.

$$\log WS = 5.09 - 1.28 \times \log Koc \qquad [2] \text{ eq. 11-4}$$

$$\log Kow = -1.07 + 1.358 \times \log Koc \qquad [2] \text{ eq. 11-5}$$

$$\log BCF_{(f)} = -1.579 + 1.119 \times \log Koc \qquad [1] \text{ eq. 11-6}$$

$$\log BCF_{(t)} = -2.024 + 1.225 \times \log Koc \qquad [1] \text{ eq. 11-7}$$

The numerical values in these equation are standard factors. Recall in Part I that the $_{(f)}$ and $_{(t)}$ after BCF are for a continuous flow and static fish study [2].

Now use these equations to solve WS, Kow, and BCFs. In Part I 2,4-D has a Koc of 20, which is equivalent to log Koc 1.301. Find the value for WS, Kow, and BCFs.

$$\log WS = 5.09 - 1.28 \log Koc$$
$$= 5.09 - 1.28 \times 1.301$$
$$= 3.424$$
$$WS = 2,658 \text{ ppm}$$

$$\log Kow = -1.07 + 1.358 \log Koc$$
$$= -1.07 + 1.358 \times 1.301$$
$$= 0.697$$
$$Kow = 4.9$$

$$\log BCF_{(f)} = -1.579 + 1.119 \log Koc$$
$$= -1.579 + 1.119 \times 1.301$$
$$= -0.123$$
$$BCF_{(f)} = 0.753$$

$$\log BCF_{(t)} = -2.024 + 1.225 \log Koc$$
$$= -2.024 + 1.225 \times 1.301$$
$$= -0.43$$
$$BCF_{(t)} = 0.371$$

Now we have the following values for 2,4-D: WS 2,654; Koc 20; Kow 4.9; $BCF_{(f)}$ 0.753; and $BCF_{(t)}$ 0.371. Using the predictive method as described in Part I and given under solubility, 2,4-D would be assessed as follows:

- WS and Koc indicates that leaching, runoff, and biodegradation could occur, and that adsorption in soil and accumulation in soil should not occur.
- Kow and BCFs indicate that bioaccumulation in the food chain should not occur; however, residues could be expected in the food chain based on values for BCFs.
- Exposure would be by consumption of contaminated food and water. Volatility could be another route of exposure; however, there are no data presented on volatility, thus it cannot be assessed if inhalation

may be another route of exposure. The aggregate, antagonistic, and synergistic effects are unknown; therefore, effects due to exposure with other hazardous chemicals cannot be assessed.

Although the values calculated here using Koc are different from those calculated by using WS, the end analysis is the same. Remember that there will always be differences and errors using these equations, but the end analysis is acceptable.

EXERCISE

Solve each of the following for WS, Kow and BCFs; write a fate and transport prediction analysis for each; and write an exposure analysis for each. Use Table 10-1, Chapter 10 under solubility to help make predictions.

#1. *Parathion:* Koc 4,800 [2]; adverse effects (possible death).
#2. *Dimilin:* Koc 6,790 [2]; potential carcinogen and potential mutagen.
#3. *Lindane:* Koc 911 [2]; a potential carcinogen, teratogen, mutagen and neurotoxin; can abort pregnancies.
#4. *Silvex:* Koc 2,600 [2]; a potential carcinogen and teratogen; can abort pregnancies.
#5. *Thiabendazole (MBC):* Koc 1,720 [2]; a potential teratogen and mutagen; reproduction hazards; hazards to wildlife.

ANSWERS

#1. *Parathion :* Koc 4,800 = log Koc 3.68; adverse effects (possibly death).

Predictive Equation Analysis

$$\log WS = 5.09 - 1.28 \log Koc$$
$$= 5.09 - 1.28 \times 3.68$$
$$= 0.378$$
$$WS = 2.39 \text{ ppm}$$

$$\log Kow = -1.07 + 1.358 \log Koc$$
$$= -1.07 + 1.358 \times 3.68$$
$$= 3.93$$
$$Kow = 8,461.36$$

$$\log BCF_{(f)} = -1.579 + 1.119 \log Koc$$
$$= -1.579 + 1.119 \times 3.68$$
$$= 2.54$$
$$BCF_{(f)} = 345.88$$

$$\log BCF_{(t)} = -2.024 + 1.225 \log Koc$$
$$= -2.024 + 1.225 \times 3.68$$
$$= 2.48$$
$$BCF_{(t)} = 304.79$$

Fate and Transport Analysis

- WS indicates that parathion should adsorb to soil, accumulate in soil, and be bioaccumulated; should not leach, runoff, or be biodegraded.
- Koc indicates that adsorption to soil should occur.
- Kow indicates that bioaccumulation should occur; however, death may occur in the organism prior to high accumulation, because it was so stated in the problem.
- BCFs indicate that residues would be present in animals and thus food chain contamination is possible. Since this chemical is an organic phosphate, some metabolism should occur in plants and animals—only an experienced person would know this.

Exposure Analysis

Residues of this chemical could contaminate the food chain; therefore, consumption of contaminated food and water should be avoided. Inhalation cannot be considered, because no data have been presented on volatility. This chemical can cause death to humans and wildlife, so due care must be taken with its use.

#2. *Dimilin:* Koc 6,790 = log Koc 3.81; a potential carcinogen and mutagen.

Predictive Equation Analysis

$$\log \text{WS} = 5.09 - 1.28 \log \text{Koc}$$
$$= 5.09 - 1.28 \times 3.82$$
$$= 0.185$$
$$\text{WS} = 1.53 \text{ ppm}$$

$$\log \text{Kow} = -1.07 + 1.358 \log \text{Koc}$$
$$= -1.07 + 1.358 \times 3.82$$
$$= 413$$
$$\text{Kow} = 12,609$$

$$\log \text{BCF}_{(f)} = -1.579 + 1.119 \log \text{Koc}$$
$$= -1.579 + 1.119 \times 3.82$$
$$= 2.709$$
$$\text{BCF}_{(f)} = 511.69$$

$$\log \text{BCF}_{(t)} = -2.024 + 1.225 \log \text{Koc}$$
$$= -2.024 + 1.225 \times 3.82$$
$$= 2.67$$
$$\text{BCF}_{(t)} = 467.95$$

Fate and Transport Analysis

- WS of Dimilin, also called diflubenzuron, indicates that it should be adsorbed to soil, be accumulated in soi,l and be bioaccumulated; should not leach, runoff, or be biodegraded.
- Koc indicates that the chemical could go either way regarding adsorption; however, WS data and Kow data indicate adsorption could occur.
- Kow indicates that bioaccumulation could occur.
- BCFs indicate that residues could be expected in the food chain.

Exposure Analysis

Residues of this chemical could contaminate the food chain. Consumption of contaminated food and water should be avoided because this chemical is a potential carcinogen and mutagen. The chemical has been intensely applied over residences, farms, people, etc., to control gypsy moths. This exposure to humans and the environment may be harmful. Inhalation is a problem when aerially sprayed over residential areas.

#3. *Lindane:* Koc 911 = log Koc 2.96; a potential carcinogen, mutagen, teratogen and neurotoxin; can abort pregnancies.

Predictive Equation Analysis

$$\log WS = 5.09 - 1.28 \log Koc$$
$$= 5.09 - 1.28 \times 2.96$$
$$= 1.3$$
$$WS = 20 \text{ ppm}$$

$$\log Kow = -1.07 + 1.358 \log Koc$$
$$= -1.07 + 1.358 \times 2.96$$
$$= 2.949$$
$$Kow = 890.59$$

$$\log BCF_{(f)} = -1.579 + 1.119 \log Koc$$
$$= -1.579 + 1.119 \times 2.96$$
$$= 1.73$$
$$BCF_{(f)} = 54.1$$

$$\log BCF_{(t)} = -2.024 + 1.225 \log Koc$$
$$= -2.024 + 1.225 \times 2.96$$
$$= 1.6$$
$$BCF_{(t)} = 39.99$$

Fate and Transport Analysis

- WS of lindane indicates that this chemical could go either way in regards to adsorption to soil, accumulation in soil, bioaccumulation, leaching, runoff, and biodegradation.
- Koc indicates this chemical should not adsorb to soil.
- Kow indicates that this chemical could go either way in regards to bioaccumulation.
- BCFs indicate that bioaccumulation in not likely; however, residues may be in the food chain.

Exposure Analysis

Because of the hazardous effects caused by this chemical, consumption of contaminated food and water could be hazardous. Inhalation cannot be assessed because data on volatility have not been presented.

#4. *Silvex:* Koc 2,600 = log Koc 3.42; a potential carcinogen and teratogen; can abort pregnancies.

Predictive Equation Analysis

$$\log WS = 5.09 - 1.28 \log Koc$$
$$= 5.09 - 1.28 \times 3.415$$
$$= 0.7188$$
$$WS = 5.23 \text{ ppm}$$

$$\log Kow = -1.07 + 1.358 \log Koc$$
$$= -1.07 + 1.358 \times 3.415$$
$$= 3.568$$
$$Kow = 3,694.62$$

$$\log BCF_{(f)} = -1.579 + 1.119 \log Koc$$
$$= -1.579 + 1.119 \times 3.415$$
$$= 2.24$$
$$BCF_{(f)} = 174.74$$

$$\log BCF_{(t)} = -2.024 + 1.225 \log Koc$$
$$= -2.024 + 1.225 \times 3.415$$
$$= 2.159$$
$$BCF_{(t)} = 144.34$$

Fate and Transport Analysis

- WS indicates that silvex should be adsorbed to soil, accumulated in soil, and bioaccumulated; should not leach, runoff, or be biodegraded.
- Koc indicates that this chemical could go either way in regards to adsorption to soil.
- Kow indicates that this chemical should bioaccumulate.
- BCFs indicate that residues could be expected in the food chain.

Exposure Analysis

Due to the toxic properties of this chemical, consumption of contaminated food and water should be avoided. Inhalation cannot be assessed because data on volatility have not been presented.

#5. *Thiabendazole (MBC):* Koc 1,720 = log Koc 3.24; a potential teratogen and mutagen; hazardous to wildlife; reproduction hazard.

Predictive Equation Analysis

$$\log WS = 5.09 - 1.28 \log Koc$$
$$= 5.09 - 1.28 \times 3.235$$
$$= 0.949$$
$$WS = 8.896 \text{ ppm}$$

$$\log Kow = -1.07 + 1.358 \log Koc$$
$$= -1.07 + 1.358 \times 3.235$$
$$= 3.323$$
$$Kow = 2,104.41$$

$$\log BCF_{(f)} = -1.579 + 1.119 \log Koc$$
$$= -1.579 + 1.119 \times 3.235$$
$$= 2.04$$
$$BCF_{(f)} = 109.89$$

$$\log BCF_{(t)} = -2.024 + 1.225 \log Koc$$
$$= -2.024 + 1.225 \times 3.235$$
$$= 1.938$$
$$BCF_{(t)} = 86.87$$

Fate and Transport Analysis

- WS of thiabendazole indicates that this chemical should be adsorbed in soil, accumulated in soil, and bioaccumulated; should not leach, runoff, or be biodegraded.
- Koc indicates that this chemical should not adsorb in soil.
- Kow indicates that this chemical should bioaccumulate.
- BCFs indicate that residues could be expected in the food chain.

Exposure Analysis

Because of the toxic properties of this chemical, consumption of contaminated food and water should be avoided. Inhalation cannot be assessed because no data have been presented on volatility.

REFERENCES

[1] Layman, Warren J., William F. Reehl and David H. Rosenblatt, Ph.D., *Handbook of Chemical Property Estimation Methods*, McGraw-Hill Book Company, 1982.
[2] Kenaga, E. E. and C. A. I. Goring, "Relationship Between Water Solubility, Soil Sorption, Octanol-Water Partitioning, and Concentration of Chemicals in Biota", *Special Technical Publication 707, American Society for Testing and Materials*, 1980.

Chapter 12
Bioconcentration and Bioaccumulation

Bioaccumulation of organic chemicals in animals is a function of a chemical's ability to become soluble in fat. When this occurs, a chemical is less prone to be metabolized by animals. Lipo-soluble (fat soluble) chemicals will bioaccumulate in fat tissue of animals and are subject to a limited amount of metabolism. A chemical of this type, will not or will only be slowly dissipated or depurated when the animal is no longer exposed to the chemical. Chemicals that are not lipo-soluble should be metabolized, dissipated, and depurated in animals. In either case, residues of the chemical can be found in an animal for some time. These residues—especially the lipo-soluble residues—may show up in mother's milk. One way to measure this is by Kow, which was previously discussed.

Another way to measure and predict if a chemical bioaccumulates (called bioconcentration factor (BCF)) is with the use of mathematical equations. WS, Koc, and Kow were previously used to calculate or predict BCFs. Bioaccumulation can be expressed as a bioconcentration factor (BCF) to determine if a chemical will bioaccumulate in an aquatic organism (i.e., fish, shale fish, crabs, etc.). The following is an actual method used to determine BCF.

$$BCF = \frac{\text{concentration of chemical in the aquatic organism (wet weight)}}{\text{concentration of chemical in water}} \qquad \text{eq. 12-1}$$

The experimental conditions are at equilibrium when carrying out this test and when using this equation. Under actual environmental conditions, equilibrium conditions would be almost impossible to maintain. The reasons being that under environmental conditions temperature, concentration of the chemical in sediment (either suspended or bottom), precipitation, etc. are uncontrollable. All these conditions and more have a bearing on the actual amount of chemical that may bioaccumulate. The aquatic organism used in these studies is usually fish. The fish are scaled, deheaded, detailed, and eviscerated for chemical analysis. Only the edible fish is considered for BCF. Both the numerator and the denominator are in the same units of measure, and these units of measure cancel, leaving the BCF value with no units.

Why is BCF important? BCF—and likewise Kow—are a measure of potential food-chain contamination, as well as for the possibility of residues of a chemical to be found in humans. BCFs are an indicator of a chemical residue in an aquatic organism; however, it can apply to any organism that has fat tissues. Once a chemical is in an organism it can be passed along the food chain and on to humans. If the residue is in the aquatic organism, then it is in water, and water may contaminate other environments that will eventually be consumed by humans and other animals.

Fish are used in BCF studies, because of ease. Fish can also take up chemical residues from water, suspended and bottom sediment, and contaminated aquatic organisms. Aquatic organisms, therefore, play a very important role in predicting bioaccumulation in the food chain. BCF also play a role in predicting WS, Koc, and Kow, when these data are not available. BCFs are another tool that can be used to make predictions on fate and transport of a chemical in the environment.

If the BCFs of a chemical are known, then equations can be used to calculate WS, Koc, and Kow. BCF is expressed as $BCF_{(f)}$ for the flow-through fish study and $BCF_{(t)}$ for the static fish study. These fish studies are a laboratory aquatic bioassay study or a model ecosystem study. The flow-through system is when fresh water is continuously added as well as the chemical. The static system uses the same water and the chemical is added only once.

The following mathematical equations use known BCF values.

$\log WS = 2.531 - 0.916 \log BCF_{(f)}$ [2] eq. 12-2

$\log Koc = 1.963 + 0.681 \log BCF_{(f)}$ [1] eq. 12-3

$\log Kow = 2.312 + 0.809 \log BCF_{(f)}$ [2] eq. 12-4

$\log BCF_{(t)} = 0.717 + 0.703 \log BCF_{(f)}$ [2] eq. 12-5

$\log WS = 1.903 - 0.69 \log BCF_{(t)}$ [2] eq. 12-6

$\log Koc = 1.886 + 0.681 \log BCF_{(t)}$ [1] eq. 12-7

$\log$ Kow $= 2.07 + 0.731 \log BCF_{(t)}$ [2] eq. 12-8

$\log BCF_{(f)} = 0.024 + 1.074 \log BCF_{(t)}$ [2] eq. 12-9

The numerical values in these equations are standard factors. Recall that $_{(f)}$ and $_{(t)}$ after the BCF stand for a continuous flow-through fish study and static fish study. Do not get these confused, or mixed, when calculating. The prediction will be the same; either the chemical bioaccumulates or not.

Now let us practice using these equations to find the unknowns WS, Koc, Kow, and a BCF.

Example: Chlorpyrifos/Dursban: $BCF_{(f)}$ 450 = log $BCF_{(f)}$ 2.653; BCF(t) 320 = log $BCF_{(t)}$ 2.505[2]; WS 0.3 ppm; Koc 13,600; Kow 97,700 [3]. The extra data are given for comparisons with the calculated values. However, please note that these values will differ and, in some instances, by large values. Equations may have a large degree of error; however, they still can be used to make predictions. In most cases, the prediction would be the same even with errors.

Use log $BCF_{(f)}$ 2.653 to find WS, Koc, Kow, and $BCF_{(t)}$.

$\log$ WS $= 2.531 - 0.916 \log BCF_{(f)}$
            $= 2.653 - 0.916 \times 2.653$
            $= 0.101$
WS $= 1.26$ ppm

$\log$ Koc $= 1.9634 + 0.681 \log BCF_{(f)}$
            $= 1.9634 + 0.681 \times 2.653$
            $= 3.77$
Koc $= 5,884$

$\log$ Kow $= 2.312 + 0.809 \log BCF_{(f)}$
            $= 2.312 + 0.0809 \times 2.653$
            $= 4.458$
Kow $= 28,726$

$\log BCF_{(t)} = 0.717 + 0.703 \log BCF_{(f)}$
            $= 0.717 + 0.703 \times 2.653$
            $= 2.582$
$BCF_{(t)}$ $= 382$

Use log $BCF_{(t)}$ 2.505 to find WS, Koc, Kow and $BCF_{(f)}$.

$$\begin{aligned}
\log WS &= 1.903 - 0.69 \log BCF_{(t)} \\
&= 1.903 - 0.69 \times 2.505 \\
&= 0.175 \\
WS &= 1.49
\end{aligned}$$

$$\begin{aligned}
\log Koc &= 1.886 + 0.681 \log BCF_{(t)} \\
&= 2.886 + 0.681 \times 2.505 \\
&= 3.59 \\
Koc &= 3,907
\end{aligned}$$

$$\begin{aligned}
\log Kow &= 2.07 + 0.731 \log BCF_{(t)} \\
&= 2.07 + 0.731 \times 2.505 \\
&= 3.901 \\
Kow &= 7,964.4
\end{aligned}$$

$$\begin{aligned}
\log BCF_{(f)} &= 0.024 + 1.074 \log BCF_{(t)} \\
&= 0.024 + 1.074 \times 2.505 \\
&= 2.714 \\
BCF_{(f)} &= 518
\end{aligned}$$

We now have a lot of data, some was given and some was calculated. Using the method given in Part I to predict fate and transport and exposure, let's group the data together to visualize just what data we have.

TYPE DATA	GIVEN	CALCULATED $BCF_{(f)}$	$BCF_{(t)}$
WS	0.3 ppm	1.26 ppm	1.49 ppm
Koc	13,600	5,884	3,907
Kow	97,700	28,726	7,964.4
$BCF_{(f)}$	450	—	518
$BCF_{(t)}$	320	382	—

WS for all indicate that chlorpyrifos should adsorb in soil, accumulate in soil, bioaccumulate, and should not leach, runoff, or be biodegraded.

Koc for the calculated values indicate that this chemical could go either way as to adsorption. The given Koc value indicates that it should be adsorbed.

Kow for all indicate that this chemical should bioaccumulate.

BCFs indicate that residues may be expected in the food chain. The values for BCFs may indicate that metabolism could occur in animals and this could mean the presence of metabolites in animals and the food chain.

Exposure should be by consumption of contaminated food and water. Inhalation cannot be assessed because no data are given on volatility. The toxic effects cannot be assessed because there are no data given.

EXERCISE

For each of the following find the WS, Koc, Kow, and BCF; write a fate and transport analysis for each; and write an exposure analysis for each. Table 12-1, Chapter 12 will be needed to help in these predictions.

#1. *Chlordane:* $BCF_{(f)}$ 11,400 [2]; a potential carcinogen.

#2. *Diazinon:* $BCF_{(f)}$ 35 [2]; kills avian species.

#3. *Endrin:* $BCF_{(f)}$ 4,050 [2]; a potential carcinogen and teratogen.

#4. *Kepone:* $BCF_{(f)}$ 8,400 [2]; a potential neurotoxin, teratogen, and carcinogen.

#5. *Toxaphene:* $BCF_{(f)}$ 26,4000 [2]; a potential teratogen, carcinogen, and mutagen; aborts births; wildlife hazards.

#6. *Dieldrin:* $BCF_{(t)}$ 4,420 [2]; a potential carcinogen; wildlife hazards.

#7. *Heptachlor:* $BCF_{(t)}$ 2,150 [2]; a potential carcinogen; wildlife hazards.

#8. *Leptophos:* $BCF_{(t)}$ 1,440 [2]; a potential neurotoxin and causer of paralysis; kills.

#9. *Mirex:* $BCF_{(t)}$ 220 [2]; a potential neurotoxin, carcinogen, and teratogen.

#10. *Picloram:* $BCF_{(t)}$ 0.02 [2]; a potential teratogen and carcinogen.

ANSWERS

#1. *Chlordane*: $BCF_{(f)}$ 11,400 = log $BCF_{(f)}$ 4.0569; carcinogen.

Predictive Equation Analysis

log WS = 2.531 − 0.916 log $BCF_{(f)}$
        = 2.531 − 0.916 × 4.0569
        = − 1.1851
WS   = 0.0653 ppm

$$\log \text{Koc} = 1.963 + 0.681 \log \text{BCF}_{(f)}$$
$$= 1.963 + 0.681 \times 4.0569$$
$$= 4.726$$
$$\text{Koc} = 53,180.069$$

$$\log \text{Kow} = 2.312 + 0.809 \log \text{BCF}_{(f)}$$
$$= 2.312 + 0.809 \times 4.0569$$
$$= 5.594$$
$$\text{Kow} = 392,673.958$$

$$\log \text{BCF}_{(t)} = 0.717 + 0.703 \log \text{BCF}_{(f)}$$
$$= 0.717 + 0.703 \times 4.569$$
$$= 3.569$$
$$\text{BCF}_{(t)} = 3,706.813$$

Fate and Transport Analysis

- WS indicates that chlordane should be adsorbed and accumulated in soil, be bioaccumulated, should not be biodegraded or leached in soil, and should not run off of soil unless with soil particles.
- Koc indicates that this chemical should be adsorbed and accumulated in soil, and be bioaccumulated.
- Kow indicates that this chemical would bioaccumulate in animals. Metabolism of this chemical in animals would probably be very slow, based on the high BCF values.
- This chemical may cause cancer and, since it should bioaccumulate in the food chain, no animal is safe from its toxic effects.

Exposure Analysis

If this chemical gets in the environment, contamination to the food chain would occur; therefore, consumption of contaminated food and water should be avoided. There are no data given on volatility, but inhalation should be avoided because this chemical is a carcinogen.

#2. Diazinon: $\text{BCF}_{(f)}$ 35 = $\log \text{BCF}_{(f)}$ 1.544; kills avian species.

Predictive Equation Analysis

$$\log \text{WS} = 2.531 - 0.916 \log \text{BCF}_{(f)}$$
$$= 2.531 - 0.916 \times 1.544$$
$$= 1.1166$$
$$\text{WS} = 13 \text{ ppm}$$

$$\log K_{oc} = 1.963 + 0.681 \; BCF_{(f)}$$
$$= 1.963 + 0.681 \times 1.544$$
$$= 3.014$$
$$K_{oc} = 1,033.865$$

$$\log K_{ow} = 2.312 + 0.809 \log BCF_{(f)}$$
$$= 2.312 + 0.809 \times 1.544$$
$$= 3.561$$
$$K_{ow} = 3,639.95$$

$$\log BCF_{(t)} = 0.717 + 0.703 \log BCF_{(f)}$$
$$= 0.717 + 0.703 \times 1.544$$
$$= 1.824$$
$$BCF_{(t)} = 63.4$$

Fate and Transport Analysis

- WS of diazinon indicates that it could go either way in regards to adsorption, biodegradation, accumulation, leaching, or run off in or from soil, and should not bioaccumulate.
- Koc indicates that this chemical should not adsorb in soil.
- Kow indicates that this chemical should not bioaccumulate.
- BCF indicates that bioaccumulation should not occur.
- This chemical is toxic to birds, therefore, extreme caution should be taken in its use as a pesticide. Since it is toxic to birds, there is the likelihood that it is toxic to other animal species, at some dose. This chemical is an organo phosphate, therefore, it is likely to be biodegraded in soil and metabolized in animals. If biodegraded and metabolized, accumulation and bioaccumulation is not likely. Although metabolism is possible, residues could be expected in organism if exposure occurred.

Exposure Analysis

If this chemical is used in the environment, bird kills could be expected. If this chemical contaminated food and water, consumption should be avoided. Inhalation should be avoided because diazinon could be hazardous to humans and other animals.

#3. *Endrin*: $BCF_{(f)}$ 4050 = $BCF_{(f)}$ 3.607; a potential teratogen and carcinogen.

Predictive Equation Analysis

$$\log WS = 2.513 - 0.916 \log BCF_{(f)}$$
$$= 2.513 - 0.916 \times 3.607$$
$$= -0.7734$$
$$WS = 0.16848 \text{ ppm}$$

$$\log Koc = 1.963 + 0.681 \log BCF_{(f)}$$
$$= 1.963 + 0.681 \times 3.607$$
$$= 4.3194$$
$$Koc = 20{,}862.53$$

$$\log Kow = 2.312 + 0.809 \log BCF_{(f)}$$
$$= 2.312 + 0.809 \times 3.607$$
$$= 5.23$$
$$Kow = 169{,}849$$

$$\log BCF_{(t)} = 0.717 + 0.703 \log BCF_{(f)}$$
$$= 0.717 + 0.703 \times 3.607$$
$$= 3.2527$$
$$BCF_{(t)} = 1{,}789$$

Fate and Transport Analysis

- WS of endrin indicates that this chemical should be adsorbed and accumulated in soil, bioaccumulated, and should not leach, run off, or be biodegraded.
- Koc indicates that this chemical should be adsorbed and accumulated in soil.
- Kow indicates that it should bioaccumulate and that food-chain contamination could occur.
- BCF values indicate that bioaccumulation should occur.
- This chemical is a carcinogen and teratogen, and can bioaccumulate in the food chain.

Exposure Analysis

If this chemical gets in the environment, contamination to the food chain could occur; therefore, consumption of contaminated food and water should be avoided. There are no data presented on volatility, but inhalation should be avoided because this chemical is a carcinogen and teratogen.

#4. *Kepone*: $BCF_{(f)}$ 8,400 = log $BCF_{(f)}$ 3.924; a potential neurotoxin, carcinogen, and teratogen.

Predictive Equation Analysis

log WS $= 2.531 - 0.916$ log $BCF_{(f)}$
$\quad\quad = 2.531 - 0.916 \times 3.924$
$\quad\quad = -1.0636$
WS $\quad = 0.08636$ ppm

log Koc $= 1.963 + 0.681$ log $BCF_{(f)}$
$\quad\quad\quad = 1.963 + 0.681 \times 3.924$
$\quad\quad\quad = 4.6352$
Koc $\quad = 43{,}176.158$

log Kow $= 2.312 + 0.809$ log $BCF_{(f)}$
$\quad\quad\quad = 2.312 + 0.809 \times 3.924$
$\quad\quad\quad = 5.4865$
Kow $\quad = 306{,}560.36$

log $BCF_{(t)} = 0.717 + 0.703$ log $BCF_{(f)}$
$\quad\quad\quad\quad = 0.717 + 0.703 \times 3.924$
$\quad\quad\quad\quad = 3.4756$
$BCF_{(t)} \quad = 2{,}989.317$

Fate and Transport Analysis

- WS of kepone indicates that it should adsorb and accumulate in soil, be bioaccumulated, and should not be biodegraded, leached in soil, or run off of soil, unless with soil particles.
- Koc indicates that this chemical should be adsorbed in soil.
- Kow indicates that it should be bioaccumulated and this should result in food-chain contamination.
- This chemical is a carcinogen, teratogen, and a neurotoxin, and should bioaccumulate in the food chain.

Exposure Analysis

If this chemical gets into the environment, contamination to the food chain would occur; therefore, consumption of contaminated food and water should be avoided. This chemical was released into the aquatic environment and aquatic species were restricted from commercial fishing and sports fishers were informed not to eat the fish. This restriction is over 10 years old. This chemical could be present in bottom sediment for hundreds, or even thousands, of years. Although there are no data on volatility, inhalation should be avoided, due to kepone's numerous toxic effects.

#5. *Toxaphene:* $BCF_{(f)}$ 26,400 = log $BCF_{(f)}$ 4.4216; a potential teratogen, carcinogen, and mutagen; aborted fetuses; hazardous to wildlife.

Predictive Equation Analysis

$$\log WS = 2.53 - 0.916\ BCF_{(f)}$$
$$= 2.53 - 0.916 \times 4.4216$$
$$= -1.52$$
$$WS = 0.3018\ ppm$$

$$\log K_{oc} = 1.963 + 0.681\ \log BCF_{(f)}$$
$$= 1.963 + 0.681 \times 4.4216$$
$$= 4.974$$
$$K_{oc} = 94,212.73$$

$$\log K_{ow} = 2.312 + 0.809\ \log BCF_{(f)}$$
$$= 2.312 + 0.809 \times 4.4216$$
$$= 5.889$$
$$K_{ow} = 774,594.48$$

$$\log BCF_{(t)} = 0.717 + 0.703\ \log BCF_{(f)}$$
$$= 0.717 + 0.703 \times 4.4216$$
$$= 3.8254$$
$$BCF_{(t)} = 6,689.36$$

Fate and Transport Analysis

- WS of toxaphene indicates that it should be adsorbed and accumulated in soil, bioaccumulated, should not be biodegraded or leached in soil, and should not run off of soil, unless with soil particles.
- K_{oc} indicates that this chemical should not be adsorbed in soil.
- K_{ow} indicates that it should bioaccumulate and could contaminate the food chain.
- BCF values indicate that this chemical should bioaccumulate and could cause food-chain contamination.
- This chemical has many toxic properties and can bioaccumulate and cause contamination in the food chain.

Exposure Analysis

If this chemical gets into the environment it could contaminate the food chain and cause death to fetuses. Consumption of contaminated water and food should be avoided. There are no data on volatility; however, inhalation of this chemical should be avoided due to the numerous toxic effects of toxaphene.

#6. *Dieldrin*: $BCF_{(t)}$ 4,420 = log $BCF_{(t)}$ 3.6454; a potential carcinogen; hazardous to wildlife.

Predictive Equation Analysis

log WS	$= 1.903 - 0.69 \log BCF_{(t)}$
	$= 1.903 - 0.69 \times 3.6454$
	$= -0.612$
WS	$= 0.2441$ ppm

log Koc	$= 1.886 + 0.681 \log BCF_{(t)}$
	$= 1.886 + 0.681 \times 3.6454$
	$= 4.3685$
Koc	$= 23,362.397$

log Kow	$= 2.07 + 0.731 \log BCF_{(t)}$
	$= 2.07 + 0.731 \times 3.6454$
	$= 4.7348$
Kow	$= 54,298.446$

log $BCF_{(f)}$	$= 0.024 + 1.074 \log BCF_{(t)}$
	$= 0.024 + 1.074 \times 3.6454$
	$= 3.93915$
$BCF_{(f)}$	$= 8,692.798$

Fate and Transport Analysis

- WS of dieldrin indicates that this chemical should adsorb and accumulate in soil, be bioaccumulated, and should not be biodegraded, leached, or run off, unless with soil particles.
- Koc indicates that this chemical should be adsorbed in soil.
- Kow indicates that it should be bioaccumulated and could cause food-chain contamination.
- BCF indicates that bioaccumulation should occur.

- This chemical is hazardous to wildlife and should not be released or put into the environment. It is also a carcinogen and exposure to animals should be prevented.

Exposure Analysis

If this chemical gets into the environment, avoid consumption of contaminated food and water. Wildlife are in danger if this chemical is released into the environment. No data are presented to discern if this chemical is volatile. Inhalation should be avoided, due to the carcinogenic effects of dieldrin.

#7. *Heptachlor:* $BCF_{(t)}$ 2,150 = log $BCF_{(t)}$ 3.3324; a potential carcinogen; hazardous to wildlife.

Predictive Equation Analysis

$$\log WS = 1.903 - 0.69 \log BCF_{(t)}$$
$$= 1.903 - 0.69 \times 3.3324$$
$$= -0.3964$$
$$WS = 0.4015$$

$$\log Koc = 1.886 + 0.681 \log BCF_{(t)}$$
$$= 1.886 + 0.681 \times 3.3324$$
$$= 4.1554$$
$$Koc = 14,302.106$$

$$\log BCF_{(f)} = 0.024 + 1.074 \log BCF_{(t)}$$
$$= 0.024 + 1.074 \times 3.3324$$
$$= 3.60299$$
$$BCF_{(f)} = 4,008.645$$

Fate and Transport Analysis

- WS indicates that heptachlor should be adsorbed and accumulated in soil, bioaccumulated, and should not be biodegraded, leached, or run off, unless with soil particles.
- Koc indicates that this chemical should be adsorbed in soil.
- Kow indicates that it should be bioaccumulated and could cause food-chain contamination if it is released into the environment.
- BCF indicates bioaccumulation and potential for food-chain contamination.

- This chemical is hazardous to wildlife and is carcinogenic; therefore, it should not be released into the environment.

Exposure Analysis

Heptachlor is a carcinogen, hazardous to wildlife and can bioaccumulate in the food chain; therefore, consumption of contaminated food and water should be avoided. No data are presented on volatility; however, this chemical is a carcinogen and its inhalation should be avoided.

#8. *Leptophos*: $BCF_{(t)}$ 1,440 = log $BCF_{(t)}$ 3.1505; a potential carcinogen and causer of paralysis; kills.

Predictive Equation Analysis

$$
\begin{aligned}
\log WS &= 1.903 - 0.69 \log BCF_{(t)} \\
&= 1.903 - 0.69 \times 3.1504 \\
&= -0.276 \\
WS &= 0.5293 \text{ ppm}
\end{aligned}
$$

$$
\begin{aligned}
\log Koc &= 1.886 + 0.681 \log BCF_{(t)} \\
&= 1.886 + 0.681 \times 3.1584 \\
&= 4.0369 \\
Koc &= 10{,}886.5189
\end{aligned}
$$

$$
\begin{aligned}
\log Kow &= 2.07 + 0.731 \log BCF_{(t)} \\
&= 2.07 + 0.731 \times 3.1584 \\
&= 4.3788 \\
Kow &= 23{,}921.609
\end{aligned}
$$

$$
\begin{aligned}
\log BCF_{(f)} &= 0.024 + 1.074 \log BCF_{(t)} \\
&= 0.024 - 1.074 \times 3.1584 \\
&= 3.41612 \\
BCF_{(f)} &= 2{,}606.88
\end{aligned}
$$

Fate and Transport Analysis

- WS of leptophos indicates that it should be adsorbed and accumulated in soil, bioaccumulated, and should not leach or run off soil, unless with soil particles, and it should not be biodegraded.
- Koc indicates that the chemical should adsorb and accumulate in soil.

- Kow indicates that the chemical should bioaccumulate and could cause food-chain contamination.
- BCF indicates bioaccumulation.
- If this chemical is released into the environment it could cause neurotoxic effects, paralysis, or death in animals.

Exposure Analysis

Eating or drinking contaminated food or water could cause adverse effects in animals, including humans. There are no data presented on volatility; however, inhalation should be prevented because of the toxic effects caused by this chemical.

#9. *Mirex:* $BCF_{(t)}$ 220 = log $BCF_{(t)}$ 2.2342; a potential neurotoxin, teratogen, and carcinogen.

Predictive Equation Analysis

$$\log WS = 1.903 - 0.69 \log BCF_{(t)}$$
$$= 1.903 - 0.69 \times 2.2342$$
$$= 0.3614$$
$$WS = 2.2983$$

$$\log Koc = 1.886 + 0.681 \log BCF_{(t)}$$
$$= 1.886 + 0.681 \times 2.2342$$
$$= 3.407$$
$$Koc = 2{,}555.584$$

$$\log Kow = 2.07 + 0.731 \log BCF_{(t)}$$
$$= 2.07 + 0.731 \times 2.2342$$
$$= 3.7032$$
$$Kow = 5{,}048.94$$

$$\log BCF_{(f)} = 0.024 + 1.074 \log BCF_{(t)}$$
$$= 0.024 + 1.074 \times 2.2342$$
$$= 2.4353$$
$$BCF_{(f)} = 265.17$$

Fate and Transport Analysis

- WS of mirex indicates that it should be adsorbed and accumulated in soil, bioaccumulated, and should not leach or run off from soil, unless with soil particles, and should not be biodegraded.
- Koc indicates that the chemical could go either way in regard to adsorption and accumulation in soil.
- Kow indicates that the chemical should bioaccumulate, causing food-chain contamination.
- BCF indicates the possibility of bioaccumulation.
- If this chemical contaminates food or water, neurotoxic, teratogenic, and carcinogenic effects could occur.

Exposure Analysis

Eating or drinking contaminated food or water could result in adverse effects to animals, including humans. Although no data are presented on volatility, inhalation should be prevented due to the numerous toxic effects of mirex.

#10. *Picloram:* $BCF_{(t)}$ $0.02 = \log BCF_{(t)} - 1.6989$; a potential teratogen and carcinogen.

Predictive Equation Analysis

$$\log WS = 1.903 - 0.69 \log BCF_{(t)}$$
$$= 1.903 - 0.69 \times -1.6989$$
$$= 3.075$$
$$WS = 1{,}189.1619$$

$$\log Koc = 1.886 + 0.681 \log BCF_{(t)}$$
$$= 1.886 + 0.681 \times -1.6989$$
$$= 0.729$$
$$Koc = 5.35857$$

$$\log Kow = 2.07 + 0.731 \log BCF_{(t)}$$
$$= 2.07 + 0.731 \times -1.6989$$
$$= 0.8281$$
$$Kow = 6.7314$$

$$\log BCF_{(f)} = 0.024 + 1.074 \log BCF_{(t)}$$
$$= 0.024 + 1.074 \times -1.6989$$
$$= -1.8006$$
$$BCF_{(f)} = 0.15826$$

Fate and Transport Analysis

- WS of picloram indicates that it should leach, run off, and be biodegraded in soil; should not be adsorbed or accumulated in soil, or bioaccumulated.
- Koc indicates that the chemical should not be adsorbed in soil, and that it should be mobile in soil.
- Kow indicates that bioaccumulation should not occur.
- BCF indicates that bioaccumulation should not occur.
- This chemical is mobile and could cause food-chain contamination. It may be biodegraded in soil, and metabolized in plants and animals; however, without knowing the rate or time it takes to reach a half-life, this can not be considered.

Exposure Analysis

This chemical is a teratogen and carcinogen and drinking of contaminated food or water should be avoided. Although no data are presented on volatility, inhalation of this chemical should be avoided.

REFERENCES

[1] Layman, Warren J., William F. Reehl and David H. Rosenblatt, Ph.D., *Handbook of Chemical Property Estimation Methods*, McGraw-Hill Book Company, 1982.
[2] Kenaga, E. E. and C. A. I. Goring, "Relationship Between Water Solubility, Soil Sorption, Octanol-Water Partitioning, and Concentration of Chemicals in Biota", *Special Technical Publication 707, American Society for Testing and Materials*, 1980.

Chapter 13
Pesticide Use Predictions

A rating and scoring system has been devised by Jerry A. Moore, Ph.D., that indicates the relative hazards of a pesticide. Dr. Moore's system can be used to compare different pesticide uses and determine the environmental hazards associated with a pesticidal use. Actual data are needed for this predictive technique. This technique is presented for those who have data or can get data to assess hazards from a pesticide use. For others, it will give a knowledge that few have and is good for information purposes and understanding.

There are twelve parameters used in Moore's rating technique; they are use patterns, fish LC 50, bird LC 50, mammal LD 50, bird LD 50, amount of pesticide used, acceptable tolerance/residue levels, percent wildlife mortality, type pesticide application, number of pesticide applications, site of application, and half life in soil.

By using these parameters, the relative hazards of a pesticide and its use—and perhaps its misuse—can be scored. A hazard index may be determined on pesticides, so that the informed users may determine if they really want to use a pesticide or not. The same hazard index may be used to determine the wildlife safety and be extrapolated for human safety analysis.

Each item in the parameters are scored from 1 to 10. The score for all are totaled and the total is divided by the number of parameters (12). The hazard index will then be between 1 and 10. The closer the number to 1 the more hazardous and, likewise, the closer the number to 10 the less hazardous.

The predictive assessment does not consider toxin effects, such as teratogenic, carcinogenic, mutagenic, etc. The predictive technique does not consider volatility, which would have an effect on animals and humans. One has to be very careful when using soil's half-life data, because residues bound in soil and any break down products should be considered. All and all, the technique can be used to formulate a hazard index for determining if you want to use a pesticide or have a particular pesticide used. This technique does not consider the aggregate, antagonistic, and synergistic effects of pesticides, nor any other chemical for which animals and humans can be exposed. These effects are not considered anyway; however, they should be if we are to be protected from hazardous chemicals.

The following twelve rating system tables are given in full with the permission of Dr. Jerry A Moore [1].

Table I.   Rating System
CROP AREA

AREA	RATING
Aquatic	1.0
Forest and Non Agricultural	2.5
Ornamental	5.0
Agricultural Areas	7.5
Home Use	10.5

Table II.   Rating System
FISH

LC 50 PARTS PER MILLION	RATING
0.000001 - 0.000009	1
0.00001 - 0.00009	2
0.0001 - 0.0009	3
0.001 - 0.009	4
0.01 - 0.09	5
0.1 - 0.9	6
1.0 - 9.9	7
10.0 - 99.9	8
100.0 - 999.9	9
1,000.0 - 10,999.9	10

Table III.   Rating System
BIRDS

LC 50 PARTS PER MILLION	RATING
1.0 - 24.9	1
25.9 - 49.9	2
50.0 - 99.9	3
100.0 - 199.9	4
200.0 - 399.9	5
400.0 - 799.9	6
800.0 - 1,599.9	7
1,600.0 - 3,199.9	8
3,200.0 - 6,399.9	9
6,400.0 - 12,799.0	10

Table IV.   Rating System
MAMMALS

LC 50 MG/KG BODY WEIGHT	RATING
0.001 - 0.009	1
0.01 - 0.09	2
0.1 - 0.9	3
1.0 - 9.9	4
10.0 - 99.9	5
100.0 - 999.9	6
1,000.0 - 1,999.9	7
2,000.0 - 2,999.0	8
3,000.0 - 3,999.9	9
4,000.0 - 4,999.9	10

Table V.   Rating System
BIRDS

LC 50 MG/KG BODY WEIGHT	RATING
0.1 - 0.9	1
1.0 - 6.24	2
6.25 - 12.4	3
12.5 - 24.9	4
25.0 - 49.9	5
50.0 - 99.9	6
100.0 - 199.9	7
200.0 - 399.9	8
400.0 - 799.9	9
800.0 - 1,599.9	10

Table VI.   Rating System
RATE

OUNCES PER ACRE	RATING
512	1
256	2
128	3
64	4
32	5
16	6
8	7
4	8
2	9
1	10

Table VII.   Rating System
TOLERANCE

TOLERANCE ALLOWED IN FOOD OR FEED PARTS PER MILLION	RATING
9	1
8	2
7	3
6	4
5	5
4	6
3	7
2	8
1	9
0	10

Table VIII.   Rating System
FIELD MORTALITY

PERCENT MORTALITY	RATING
91 - 100	1
81 - 90	2
71 - 80	3
61 - 70	4
51 - 60	5
41 - 50	6
31 - 40	7
21 - 30	8
11 - 20	9
0 - 10	10

Table IX.   Rating System
MODE AND PLACEMENT IN ENVIRONMENT

PARAMETER	RATING
ULV Aerial	1
LV Aerial	2
EC Aerial	3
WP or Dust Aerial	4
ULV Ground	5
LV Ground	6
EC Ground	7
WP or Dust Ground	8
G on top of soil by air or ground equipment	9
G below soil surface	10

Table X.   Rating System
APPLICATION AND REPLICATION

NUMBER AND TIMES	RATING
Daily	1
Twice in one week	2
Once per week	3
Once every two weeks	4
Once every four weeks	5
Once every eight weeks	6
Once every sixteen weeks	7
Once every thirty two weeks	8
Once every fifty two weeks	9
Once every two years	10

Table XI.   Rating System
SITES

SITE AND SCORE	RATING
Aquatic areas	1
Short grass	2
Leafy crops (non row)	3
Leafy crops (in rows)	4
Tall grass	5
Forage (i.e., alfalfa, clover)	6
Shrubs (herb and woody)	7
Plants bearing pods	8
Orchards	9
Forest areas	10

Table XII.   Rating System
HALF LIFE

HALF LIFE IN DAYS	RATING
512	1
256	2
128	3
64	4
32	5
16	6
8	7
4	8
2	9
1	10

The next nine tables represent three chemicals (A, B, and C). Chemical A is presented with four sets of data, chemical B with three sets of data, and chemical C with two sets of data. By varying the data it is easier to visualize how the hazard index changes. The data are real; however, the chemicals are hypothetical.

The following tables are given in full with the permission of Dr. Jerry A Moore. (1)

Table 1. Hazard index for certain variables Chemical A (4.8)

PARAMETER	SCORE
Crop Area	1.0
Fish LC 50	7.0
Bird LC 50	1.0
Mammal LD 50	4.0
Bird LD 50	2.0
Ounces/Acre	5.0
Tolerances	5.0
Field Mortality	10.0
Placement	4.0
Application/Replication	8.0
Site	1.0
Half life	7.0
Hazard index	**4.83**

Table 2. Hazard index for certain variables Chemical A (5.5)

PARAMETER	SCORE
Crop Area	10.0
Fish LC 50	7
Bird LC 50	1
Mammal LD 50	4
Bird LD 50	2
Ounces/Acre	5
Tolerances	5
Field Mortality	10.0
Placement	7
Application/Replication	8
Site	1
Half life	7
Hazard index	**5.583**

Table 3.   Hazard index for certain variables Chemical A (5.1)

PARAMETER	SCORE
Crop Area	5
Fish LC 50	7
Bird LC 50	1
Mammal LD 50	4
Bird LD 50	2
Ounces/Acre	5
Tolerances	5
Field Mortality	10.0
Placement	7
Application/Replication	8
Site	1
Half life	7
Hazard index	**5.166**

Table 4.   Hazard index for certain variables Chemical A (5.4)

PARAMETER	SCORE
Crop Area	5.0
Fish LC 50	7
Bird LC 50	1
Mammal LD 50	4
Bird LD 50	2
Ounces/Acre	5
Tolerances	5
Field Mortality	10.0
Placement	10.0
Application/Replication	8
Site	1
Half life	7
Hazard index	**5.41**

Table 5.   Hazard index for certain variables Chemical B (7.0)

PARAMETER	SCORE
Crop Area	10.0
Fish LC 50	9
Bird LC 50	9
Mammal LD 50	8
Bird LD 50	7
Ounces/Acre	5
Tolerances	8
Field Mortality	7
Placement	6
Application/Replication	4
Site	10
Half life	1
Hazard index	**7.0**

Table 6. Hazard index for certain variables Chemical B (5.5)

PARAMETER	SCORE
Crop Area	10
Fish LC 50	9
Bird LC 50	9
Mammal LD 50	8
Bird LD 50	7
Ounces/Acre	1
Tolerances	2
Field Mortality	7
Placement	2
Application/Replication	1
Site	9
Half life	1
Hazard index	**5.5**

Table 7. Hazard index for certain variables Chemical B (5.5)

PARAMETER	SCORE
Crop Area	10
Fish LC 50	9
Bird LC 50	9
Mammal LD 50	8
Bird LD 50	7
Ounces/Acre	3
Tolerances	0
Field Mortality	7
Placement	2
Application/Replication	1
Site	9
Half life	1
Hazard index	**5.5**

Table 8. Hazard index for certain variables Chemical B (4.8)

PARAMETER	SCORE
Crop Area	10
Fish LC 50	5
Bird LC 50	5
Mammal LD 50	5
Bird LD 50	5
Ounces/Acre	5
Tolerances	5
Field Mortality	5
Placement	5
Application/Replication	9
Site	1
Half life	7
Hazard index	**4.83**

Table 9.   Hazard index for certain variables Chemical C (5.2)

PARAMETER	SCORE
Crop Area	10
Fish LC 50	5
Bird LC 50	5
Mammal LD 50	5
Bird LD 50	5
Ounces/Acre	5
Tolerances	5
Field Mortality	5
Placement	5
Application/Replication	2
Site	4
Half life	7
Hazard index	**5.25**

To my knowledge, this technique has not been used to support or deny a pesticide registration by the EPA. It has not been used to formulate label restrictions on pesticide labels, nor has it been presented to the EPA or the general public for use or validation.

EXERCISE

Given the following *made-up* data, find the ratings for the two sets of data.

PARAMETER	USE TO DETERMINE RATINGS
Crop Area	aquatic & home
Ounces/ Acre	32 & 32
Tolerances	1 ppm & 0
Field Mortality	45% & 45%
Mode & Placement in Environment	ULV Aerial & EC Ground
Application/Replication	1 oz/32 wks & 1 oz/32 wks
Sites	Aquatic & Short grass
Half life	128 days & 128 days

For each of the following chemicals, data are given to find ratings for each.

CHEMICAL	LC 50 PPM FISH	LC 50 PPM BIRDS	LD 50 MG/KG MAMMALS	LD 50 MG/KG BIRDS
Carbaryl	500	1,500	151	100
Chlordane	0.01	>5,000	540	2,179
Dichlorvos	1	>5,000	80	7.8
2,4-D	50	>5,000	666	>1,000
Toxaphene	0.0028	834	69	70.7

For each chemical determine the hazard index using the ratings found. There should be two hazard indexes per chemical. Write a short predictive analysis for each.

ANSWERS

PARAMETERS	RATINGS				
	CARB-ARYL	CHLOR-DANE	DICHL-ORVOS	2,4-D	TOXAP-HENE
Crop Area	1 & 10	1 & 10	1 & 10	1 & 10	1 & 10
Fish LC 50 ppm	9 & 9	5 & 5	7 & 7	8 & 8	4 & 4
Bird LC 50 ppm 7 & 7	8 & 8	9 & 9	9 & 9	9 & 9	
Mammal LD 50 mg/kg	6 & 6	6 & 6	5 & 5	6 & 6	5 & 5
Bird LD 50 mg/kg	7 & 7	10 & 10	3 & 3	10 & 10	6 & 6
Ounces/Acre	5 & 5	5 & 5	5 & 5	5 & 5	5 & 5
Tolerances 9 & 10		9 & 10	9 & 10	9 & 10	9 & 10
Field Mortality	6 & 6	2 & 6	6 & 6	6 & 6	6 & 6
Placement 1 & 7		1 & 7	1 & 7	1 & 7	1 & 7
Application/ replication	8 & 8	8 & 8	8 & 8	8 & 8	8 & 8
Site	1 & 2	1 & 2	1 & 2	1 & 2	1 & 2
Half life	3 & 3	3 & 3	3 & 3	3 & 3	3 & 3
TOTAL	64 & 81	64 & 81	58 & 75	67 & 84	56 &73
divided by 12 Hazard Index	5.3 & 6.75	5.3 & 6.75	4.8 & 6.25	5.58 & 7	4.66 & 6.08

The hazard index is acceptable for all uses given the data. If the aquatic use of toxaphene changes, a closer look will be needed.

Remember, the above uses are only acceptable based on data given. But what if the chemicals bioaccumulate in the food chain, are volatile, cause other toxic effects (i.e., carcinogenic, mutagenic, teratogenic, oncogenic, etc.), kill bees, cause fetal abortions, or the like? Would the use of these chemicals be acceptable if data like this was given?

EXERCISE

For each chemical, consider the following data and render another environmental prediction. Pay close attention to how the prediction will change when the conditions change.

#1. *Carbaryl:* carcinogen; kills bees; found in air of homes; does not bioaccumulate.

#2. *Chlordane:* carcinogen; causes adverse effects; bioaccumulates; found in air of homes.

#3. *Dichlorvos:* oncogenic; volatile.

#4. *2,4-D:* teratogen; carcinogen; does not bioaccumulate; excreted in cow's milk.

#5. *Toxaphene:* bioaccumulates; carcinogen; aborts fetus; harmful to wildlife.

ANSWERS

#1. *Carbaryl:* Carbaryl does not bioaccumulate; however, aquatic uses could result in food-chain contamination. This contamination could result in carcinogenic effects to animals and humans consuming contaminated food and water. Uses around homes could cause bee kills, resulting in reduced pollination in the area. Carbaryl can be found in air of homes; therefore, inhalation of a carcinogen could be a problem. The aquatic and home uses should be avoided based on the data given.

#2. *Chlordane:* Chlordane can bioaccumulate in the food chain and could cause adverse effects on wildlife. Since chlordane is a carcinogen, any consumption of contaminated food or water should be avoided. Chlordane has been found in the air of homes; therefore, inhalation should be avoided. The aquatic and home uses should be avoided, based on the data.

#3. *Dichlorvos:* Dichlorvos is an oncogen and is volatile; therefore, consumption of contaminated food or water should be avoided. Inhalation should also be avoided. The aquatic and home uses should be avoided, based on the data.

#4. *2,4-D:* 2,4-D does not bioaccumulate; however, residues may be found in the food chain. This chemical is a teratogen and carcinogen, and can be excreted in milk; therefore, consumption of contaminated water or food should be avoided. There are no data on volatility; however, inhalation should be prevented. The aquatic and home uses should be avoided, based on the data.

#5. *Toxaphene:* Toxaphene is a carcinogen, mutagen, fetal aborter, hazardous to wildlife, and can bioaccumulate. Do not consume contaminated food or water, and do not inhale this chemical. The aquatic and home uses, should be avoided based on the data.

REFERENCES

[1] Moore, Jerry Arnold, "A New Concept in Pesticide Labeling," *A Dissertation Presented to the Faculty of Heed University, Heed University, Hollywood, Florida, 1980.*

Chapter 14
Similar Chemical Structures

Do some toxic chemicals have similar structures? Can these similarities be used to predict potential toxic effects, if a toxic effect is known on any one or more? The answer to both is yes; analogies may be made. However, real data is needed to discern real toxic effects. The information on such predictions is useful, in that it is a good indication that toxic effects studies should be done. You would not want to be exposed to such chemicals without really knowing if they were toxic. Examples of some groups of chemicals are given below.

ALDEHYDE

Aldehydes are a group of chemicals that have a -CHO- group in their chemical structure. By recognizing aldehyde in a chemical name, you can predict that the chemical may have adverse effects on animals, including humans, because it is a carcinogen, mutagen, oncogen, etc. Note the following chemicals:

- 2,3:4,5-Bis(2-butylene) tetrahydro-2-furaldehyde—an oncogen; causes ovarian atrophy, adverse reproductive effects, and tumors. This chemical is a pesticide.
- Acrylaldehyde\acrolein—metabolized in animals to glycidaldehyde, a carcinogen. This chemical is used in the manufacturing of perfumes, etc.
- Chloroacetaldehyde—a mutagen. This chemical is used to remove bark from trees.
- Formaldehyde—a carcinogen in laboratory animals; degrades to phosgene, a poisonous gas. This chemical is a solvent and has been used as insulation in homes.

ANILIDE, AMINE, AMINO, and AMIDE

This group of chemicals may be recognized by the group anilide, amine, amino, or amide which appears in the chemical name. This group of chemicals contain nitrogen and hydrogen in their chemical structure (NH). The following examples are given for analogies.

Anilide

- 3,4',5-Tribromosalicylanilide—a suspected carcinogen. Found in soaps, etc.
- 3,4,4'-Trichlorocarbanilide—found in detergents. A hazard? You could predict yes based on the anilide which appears in the name.
- Alachlor (2-Chloro-2',6'-diethyl-N-(methoxymethyl) acetanilide) also known as Lasso—a potential oncogen. This chemical is a pesticide.

Amine

- beta-Naphthylamine—a carcinogen which has caused bladder cancer.
- Triethanolamine (TEA)—carcinogen in laboratory animals. Found in cosmetics, cleaners, detergents, etc.
- Toluene-2,4-diamine/diaminotoluene—a carcinogen. Found in dyes, etc.
- Triethylene tetramine (TETA)—a teratogen in laboratory animals.

Amino

- 2-Acetylaminofluorene (AAF)—a carcinogen to liver and bladder and a cytotoxic teratogen. A metabolite is formed in animals and is N-Hydroxy AAF—a carcinogen. This is a pesticide.
- para-Aminophenol—a teratogen is laboratory animals and causes central nervous system depressions (CNS-dep.). Used in hair dyes,etc.
- 4-Dimethylaminobenzene—a carcinogen in animals. Used in wax products, etc.
- para-Aminobenzene—a suspected carcinogen. A pesticide, used in dyes (aniline yellow), etc.
- para-Aminobenzene hydrochloride—a suspected carcinogen. Used in dyes, lacquers, etc.
- Chlorambucil (4-(p-[bis(2-chloroethyl) amino] phenyl) butyric acid)—a carcinogen. A pesticide.

- ortho-Aminotoluene/ortho-toluene—a carcinogen. Used in textile paints, dyes, etc.
- Daminozide (Butanedioic acid mono (2,2-dimethylhydrazide) also known as B-Nine, Alar, etc.—a carcinogen. A pesticide.

Amide

- Acrylamide—a carcinogen and suspected carcinogen in humans. Used as waterproofing, soil stabilizer, etc.
- Acetamide/acetic acid amine—potential carcinogen. Used in lacquers, soldering flux, etc.
- Amitraz (N'-(2,4-Dimethylphenyl)-N-[[(2,4-dimethylphenyl) imino) methyl]-N-methylmethanimidamide) also known as BAAM, Trazid, etc.—a pesticide; a potential oncogen.
- Chlordimeform (N'-(4-Chloro-o-tolyl)-N,N-dimethylformamidine) also known as Galecron, Fundal, etc.—a potential oncogen. A pesticide.

ANILINE

This group of chemicals has NH in the structure; however, it is even of greater importance because these chemicals have the ability to form nitroso-amine/anilines, a known carcinogen. Several chemicals are given below.

- Aniline—a carcinogen and is transformed to azo compounds, a carcinogen. A pesticide, used in dyes, etc.
- Carboxin (2,3-Dihydro-5-carboxanilido-6-methyl-1,4-oxathin—may form aniline and azobenzenes. A pesticide.
- Dichlozoline also [3-(3,5-Dichlorophenyl)-5,5-dimethyl-2,4-oxazolidinedione]—may form dicholoraniline and tetracholoroazobenzene, a carcinogen. A pesticide.
- Propanil also 3',4'-Dichloropropionanilide—may form dichloroaniline and tetrachloroazobenzene, a carcinogen. A pesticide.
- para-Aminoazobenzen hydrochloride—a carcinogen. Used in dyes, lacquers, etc.

AZO and NITROSOANILINE

Organic chemical containing anilines in their structure can be degraded, metabolized, or better transformed to azobenzenes or nitrosoanilines, which are known carcinogens. Some examples are given above and several others are now given.

- N-sec-Butyl-4-tert-butyl-2,6-dinitroaniline—transformed to nitrosaniline. A pesticide.
- Botran (2,6-Dichloro-4-nitroaniline)—possible to nitrosoaniline. A pesticide.
- Basalin [N-(2-Chloroethyl)-N-propyl-2,6-dinitro-4-trifluoro-methylaniline]—may be to nitrosoanline. A pesticide.

BENZENE

- Benzene—a carcinogen and fetotoxin in laboratory animals. A pesticide, solvent, in gasoline, etc.
- para-Dichlorobenzene—a carcinogen in laboratory animals. Transformed by metabolism to form 2,5-dichlorophenol, a carcinogen. A pesticide (i.e., moth repellent).
- Phenol—fetotoxic in laboratory animals. A pesticide, antiseptic, etc.

IMIDAZOLE

Methyl N-(2-benzimidazolyl) carbamate (MBC) is a known carcinogen. Some chemicals that break down to, or could break down to, MBC are listed.

- Benomyl (Methyl N-(N-butylcarbamoyl)-2-benzimidazolyl carbamate)—a mutagen, teratogen, and can cause reproductive problems in wildlife. This chemical is a pesticide and is used as an oxidizer in sewage treatment facilities. Is it discharged in the environment from such facilities? This chemical degrades to MBC.
- Thiabendazole (2-(4'-Thiazolyl) benzimidazole) (TBC).
- Methabenzithiazuron [1-(2-Benzothiazolyl)-2,3-dimethylurea]. This chemical has two links that could cause problems—the benzimidazole and dinitroaniline group (discussed later)—which may also cause cancer.

BIPHENYL/DIPHENYL

Biphenyl are usually recognized as polychloronated biphenyls (PCBs). Both biphenyls and diphenyls can cause cancer and other side effects. Below is a list of some prospects.

- Benzidine (1,1'-Biphenyl-4,4'-diamine)—a carcinogen which causes bladder cancer. Used in the manufacturing of dyestuff.
- 3,3'-Dichlorobenzidine (DCB)—a carcinogen.
- Biphenyl—a carcinogen; and causes CNS-damage. A pesticide, heat transfer agent, etc.
- Chlorodiphenyl—a carcinogen.
- 4-Aminodiphenyl—a carcinogen to liver and kidney; tumors found in mammary glands. A rubber antioxidant.
- 4-Nitrobiphenyl—a carcinogen which has effected the bladder. A wood preservative.
- ortho and para-Aminobiphenyl—a carcinogen
- Dichloro diphenyl trichloroethane (DDT)—a carcinogen; hazardous to bird shells, etc. A pesticide.
- Methoxychlor, technical (2,2-bis(para-Methoxyphenyl)-1,1,1-trichloroethane) and related compounds—a carcinogen; hazardous to wildlife. A pesticide, also known as methoxy DDT.
- Dichloro diphenyl dichloroethane (DDD). A metabolite/degradate of DDT.
- Kelthane (1,1-bis(Chlorophenyl)-2,2,2-trichloroethane). A pesticide.
- Ethoxyclor. A pesticide.

THIOCARBAMATE

Ethylenethiourea (ETU), also known as 2-imidazolidinethione, NHCH2CH2NHCS. ETU is a degradation product and metabolite of many pesticides. There is strong evidence that ETU causes cancer in humans. Ethylenebisdithiocarbamate (EBDC) and ETU cause cancer, birth defects, genetic mutations, thyroid effects, and are acutely toxic. Knowing this information, note the similarities with the following chemicals and you be the judge if you want to be exposed to them.

- Maneb (Manganese ethylenebisdithiocarbamate). A pesticide.

- Zineb (Zinc ethylenebisdithiocarbamate). A pesticide.
- 2-Chloroallyl diethyldithiocarbamate (CDEC). A pesticide.
- Ferbam (Ferric dimethyldithiocarbamate). This chemical has caused blindness in fish, fin erosion, and death. A pesticide.
- Nabam (Disodium ethylenebisdithiocarbamate). This chemical is volatile and breaks down to carbonyl sulfide, which is flammable and highly toxic. A pesticide.
- Vapam (Sodium methylenedithiocarbamate). A pesticide.
- Asulan (methyl sulfanilylcarbamate). A potential oncogen.
- S-(2,3-Dichloroally) diisopropylthiocarbamate, also known as Diallate, Avadex, etc. A pesticide and a potential oncogen.

IMIDE and IMINE

These chemicals also have NH groups as previously discussed; however, a special listing is needed for them.

Imide

- Diethyleneimide oxide/morpholine—a carcinogen. A pesticide, solvent for waxes and dyes, etc.
- Captafol[[(cis-N-(1,1,2,2-Tetrachloroethyl) thio]-4-cyclohexene-1,2-dicarboximide) also known as Difolatan, Folid, etc.—a pesticide and potential oncogen.
- Captan (N-Trichloromethylthio-4-cyclohexene-1,2-dicarboximide) also known as Orthocide, Vancide, etc.—a pesticide and potential oncogen.

Imine

- Ethylenimine—a carcinogen in laboratory animals. Used as a clarifier in effluent.
- Propylene imine—a carcinogen in animals. Used as a manufacturing ingredients.

INDOLE

Indole is a known carcinogen and is a transformation/breakdown product of many other chemicals. Some of these chemicals are listed below.

- Indigo—maybe to indole, a carcinogen, and to isatin, which may go to indole. A dye.
- Isatin (indole-2,3-dione)—maybe to indole, a carcinogen. Used in pharmaceuticals, dyestuff, etc.
- Tryptophan (indole-alpha-aminopropanic acid)—may go to indole, a carcinogen. Nutrition—amino acid.

AMIDO

- Acephate (O,S-Dimethyl acetylphosphoramidothioate) also known as Orthene—a pesticide and a potential oncogen and is moderately toxic by ingestion.

PHTHALIMIDE

This is an imide group which was previously given; however, it was thought that this group deserved a special place as you will see. Remember thalidomide? Thalidomide causes birth defects and was taken off the market. Well phthalimide is a potential precursor and a breakdown product in many pesticides. Look out for chemicals with "imide" in their name— with phthalimide. Below are a few chemicals with phthalimide in their name.

- Captan [N-Trichloromethlythio)-cyclohex-4-ene-1,2-dicarboximide— a suspected carcinogen, teratogen, mutagen, and has effects on reproduction.
- Folpet [N-(Trichloromethylthio) phthalimide.
- Imidan [O,O-Dimethyl S-phthalimidomethyl phosphorodithioate.
- N,N'-Thiobisphthalimide.

CHLOROANILINE

Chemicals that contain a benzene ring with halogen(s) and urea may break down to chloroanilines. Chloroanilines are a precursor to azobenzenes, which are known carcinogens. Usually there are two or more chlorine atoms for the aniline to be a potential carcinogen. It is possible that any similar halogen containing compound could react in the similar manner. Some of the following chemicals are potential candidates.

- Buturon (n-(4-Chlorophenyl)-N'-methyl-N'-(1-methylprop-2-ynyl) urea.
- Chlorobromuron [3-(3-Chloro-4-bromophenyl)-1-methoxy-1-methylurea].
- Diflubenzuron [1-(4-Chlorophenyl)-3-(2,6-difulorobenzoly) urea.
- Fluometuron [1,1-Dimethyl-3-(m-trifluoromethylphenyl) urea].
- Linuron [3-(1,4-Dichlorophenyl)-1-methoxy-1-methylurea].
- Monolinuron [3-(4-Chlorophenyl)-1-methoxy-1-methylurea].
- Monuron [3-(4-Chlorophenyl)-1,1-dimethylurea].
- Neburon [1-butyl-3-(4-Dichlorophenyl)-1-methylurea].

VINYL

This group of vinyl compounds (-CH2=CH-) may also be used to make predictions as to adverse effects of other similar compounds. Some are carcinogenic and others have other effects.

- Vinyl cyanide also acrylonitrile—a carcinogen in laboratory animals and suspected in humans; causes tumors in the mammary gland, GI tract, and brain.
- Vinyl bromide also bromoethene—a carcinogen in laboratory animals; CNS-depressant.
- Vinyl chloride—a human carcinogen.

HYDRAZINE

Hydrazine (-CH2-CH2-) is a carcinogen, but is it a carcinogen in other chemical structures? Can we use the hydrazine moiety to predict carcinogens? The answer to both is yes. Some of the following chemicals can be used a examples.

- Daminozide (succinic acid 2,2-dimethylhydrazide), also known as Alar—a pesticide; a carcinogen.
- Maleic hydrazide (MH)—a pesticide; an oncogen.

- Hydrazine sulfate—a pesticide; an oncogen.
- Methyl hydrazine—a solvent, etc.; a suspected oncogen.
- Phenylhydrazine; used to make pharmaceuticals, etc.; a carcinogen.

Remember that chemicals with these similar structures can be predicted to cause adverse effects. Actual testing is needed to discern the actual adverse effects. The information is useful to discern if you want to be exposed to similar chemicals. It is also useful in making presentations and arguments to support needed testing.

In general, the structural groups look like the following. Please note that the dash (-) indicates that other elements or groups may be attached.

- aldehyde; (-CHO)
- amide; (-NH2), (-CONH2) and (-=CONH2)
- amido; (-CH3CONH-)
- amine; (-NH3), (-NH2) and (NH)
- amino; (-NH2)
- anilide; (-C6H4NHCO-)
- aniline; (-C6H4NH2)
- azo; (-C6H4NNC6H4-)
- benzene; (C6H6) and (-C6H5)
- biphenyl; (-C6H4-C6H4-)
- chloroaniline; (-C6H4NHCl)
- diphenyl; (_C6H4-C6H4-)
- hydrazine; (-CH2-CH2-)
- imidazole; (-HNCHNCHCH-)
- imide; (-C6H4CO2NH)
- imine; (-CH=NH)
- indole; (-C6H3(CH2)NH)
- nitrosoaniline; (C6H4NNC6H3NH2-)
- phthalimide; (-C6H3(CO2)NH)
- thiocarbamate; (-(NH2)2CS)
- vinyl; (CH2=CH-)

Two other groups are given below because they contain (-NH-) groups which are analogist to several previous groups. These groups are urea and thiourea. They can also be used to predict hazards. Urea also carbamide; CO(NH2)2 and thiourea also thiocarbamide; (NH2)2CS. Below are examples of several chemicals.

- Diuron (3-(3,4-Dichlorophenly)-1,1-dimethylurea
- Finuron (3-Phenyl-1,1-dimethylurea)
- 1-(Naphthyl)-2-thiourea

STEROID HORMONES

Another group yet to be mentioned is the steroid hormones. Steroids are a class of lipids containing four cyclic rings. The neucleus is made of three cyclohexane rings and an cyclopentane ring. Cholesterol is the most abundant steroid in the human body and is the starting material for steroid hormone(s) synthesis.

Steroid hormones appear to be a problem when used for medical treatments. The steroid nucleus can be used to predict problems as indicated by the following list of steroids.

- Cortisol: human teratogen
- Cortisone: questionable carcinogen; experimental reproductive effects
- Dehydrostilbestrol: postmenopause treatment; suspected human carcinogen
- Diethylstilbestrol (DES): confirmed carcinogen
- Diethylstilbestrol dipalmitate: experimental teratogen; questionable carcinogen
- Diethylstilbestrol dipropionate: confirmed carcinogen; experimental teratogen
- Estriol: treatment for menopausel suspected carcinogen
- Estrone: treatment for menopause; confirmed carcinogen
- Estradiol polyester with phosphoric acid: cancer treatment; confirmed carcinogen; experimental teratogen
- Estradiol mustard: questionable carcinogen
- Estradiol dipropionate: treatment for menopause; confirmed carcinogen; experimental teratogen
- Estradiol-3-benzoate: steroid; confirmed carcinogen; experimental teratogren
- Estradiol: regulates menstrual cycle; treatment for menopause; confirmed carcinogen; experimental teratogen
- Progesterone: essential for implanting fertile ovum; regulates menstrual cycle; confirmed carcinogen

- Testosterone: promotes growth of male gental organs; confirmed carcinogen.

A few naturally occurring chemicals are given.

SAFROLES

- Safrole: in pepper; known carcinogen.
- Isosafrole: pesticide synergist, perfumery, flavor; suspected carcinogen.
- Isosafrole -n-octylsulfoxide: pesticide; questionable carcinogen.

CHRYSENES

These tetracyclic hydrocarbons can be derived from distillation of coal tar.

- 1,11-Dimethylchrysene: questionable carcinogen.
- 3-Methylchrysene: questionable carcinogen.
- 4-Methylchrysene: questionable carcinogen.
- 5-Methylchrysene: questionable carcinogen.
- Chrysene, also 1,2-benzphenanthrene: human carcinogen.

EXERCISE

Using the previously listed groups and examples, underline the group(s) in each of the following pesticide chemicals.

- O,O-Bis (para-chlorophenyl) acetimidophosphoramidothioate, also Gophacide
- Bromacil, dimethylamine salt
- Butachlor (N-(Butoxymethyl)-2-chloro-2',6'-diethylacetanilide
- Butam (2,2-Dimethyl-N-(1-methylethyl)-N-(phenylmethyl) propanamide
- Calcium ethylenebisdithiocarbamate
- N-[[(4-Chlorophenyl) thio] methyl] phthalimide

- 2-Chloro-1-(2,4,5-trichlorophenyl) vinyl dimethyl phosphate also Gardona, Rabon, etc.
- 2-Chloro-N-[[[4-trifuoromethoxy] phenyl amino] carbonyl] benzamide
- Diflubenzuron (N-[[(4-Chlorophenyl) amino] carbonyl]-2,6-difluorobenzamide also Dimilin
- 1,2-Dimethyl-5-nitroimidazole
- Diphenylamine
- Metaldehyde
- Sodium fluoaluminate
- Sodium salicylkanilide
- Zoalene (3,5-Dinitro-o-toluamide)
- 3-(4-Isopropylphenyl)-1,1-dimethylurea

Chapter 15
Dissipation in Soil

Dissipation has many meanings and many situations, as you will soon learn. The standard definition for dissipation is the time it takes for a chemical to be degraded/non-existent. Dissipation to a chemical's half-life (T-1/2) is the time it takes for a chemical to reach 1/2 of the its original amount in any given environment. This too has many meanings and many situations. A simple definition cannot really be given, because such would not be all inclusive.

We need to have a starting point; however, we must remember that a chemical compound can break down to lower molecular weight compounds or be transformed to higher molecular weight compounds. For convenience, break down, degradation, or transformation will be all inclusive in this chapter. In other words, all products formed from a parent chemical are included, as well as the parent chemical (the word residues therefore includes all).

Dissipation should not exclude any residues that may be bound in soil (adsorbed in soil). Some residues may be bound in plants and animal tissue, and this must be considered when discussing living organisms. How can you tell if a residue is adsorbed? The previous chapters explain how to predict this. Actual chemical experimentation is needed to analytically prove bound residues.

Dissipation includes all mechanisms which reduce the amount of original residue. For example, in soil this could be volatilization from soil, leaching and runoff from soil, and complete degradation to naturally occurring compounds, which is unlikely in most cases. Dilution is not a form of dissipation or degradation.

The following list will help visualize some of the mechanisms of dissipation in the environment.

Environments	Mechanisms
Air	Phototransformation
Water	Hydrolysis
Soil	Biodegradation
Plants	Metabolism
Animals	Volatilization
	Pyrolysis
	Movement

The following scenarios describe different dissipation situations.

SCENARIO 1

As a parent chemical degrades to two degradates, they could buildup to equal 100%. If this occurs, dissipation did not occur, only degradation. A is parent chemical degrading to B & C, and B to C, thus C is building up. If this T-1/2 was 60 days, then 100% of the residues would be present. This means 50% parent chemical and 50% C.

SCENARIO 2

As a parent chemical degrades to degradates, one or more degradates could buildup. A is degrading to B, C, D, E, F and so on, and B, D, E, F and so on to C, C is building up. If this T-1/2 was 60 days, then 100% of the residues are present. This means 50% of parent and 50% C.

SCENARIO 3

As the parent chemical degrades to naturally occurring chemicals, the total residues would be reduced. A to C is parent chemical degrading, and A to B, D, E and so on, are degradates building up. Other residues not shown have degraded to naturally occurring chemicals, which means they are really dissipated and this accounts for the loss. If the T-1/2 was 60 days and 100% of the residues were not there, then there is a loss of residues. A to C accounts for 50% of the residue; however, A to B, D, E, and so on, is less than 50% and this accounts for the loss.

SCENARIO 4

As parent chemical and its residues are released via volatility, leaching, runoff, etc., residues are dissipated, but not degraded. If A to C represents 50% degradation of parent chemical at 60 days and no other degradates are found. This means there is dissipation of 50% of the original amount. The 50% lost could be due to the degradation to naturally occurring chemicals, loss to volatility, leaching, or run off. Another environment could be contaminated if the loss was not to naturally occurring chemicals and this must be considered in risk assessment and exposure assessments.

Now it should be easier to understand why one definition is difficult. Dissipation is the time it takes for total residues to reach a half-life (T-1/2) of the original residues. When discussing dissipation, degradation or T-1/2 one must be specific and describe what is meant to prevent misinterpretation of data.

Chapter 16
Estimated Exposure To Fish—
Safety Factor (SF)

Did you ever wonder how to reduce the acute toxic effects that a chemical may have on fish? That is, the lethal concentration to fish based on a 24 or 96 hour LC 50 test. There are several ways to reduce the LC 50: 1) reduce the amount of chemical(s) added in water, or 2) dilution.

Of course, this we already knew. If there is a spill on the ground, spill in water, or even an application of a chemical to water, wouldn't it be great if we could predict fish exposure to a chemical and perhaps prevent fish kill? Even preventing a kill to other aquatic organisms would be more than beneficial.

Buried deep in the archives of my notes and brain over the past 35 years is such a method—more like a hypothesis—to make such predictions. I do not recall, nor have I found references to, this or a similar prediction method. The method, which I am sure I have modified, will be presented as a mathematical method.

The predictive technique is designed to rate the potential exposure of a fish to a chemical and to determine a reduced safety margin. By doing this, one could determine the margin of safety to fish from a chemical spill or intentional discharge of a chemical or a pesticide in the aquatic environment.

The predictive technique will enable the user to determine a margin of safety in an area of high or low rainfall, and high or low water runoff.

The predictive technique will enable the user to determine the margin of safety from mixing of a chemical in water or adding water to dilute the chemical to meet a safety factor (SF).

The equation for a chemical in water is the basic equation and this equation is based on fish toxicity under static conditions. The equation is the chemical amount times ppm in water divided by the LC 50 times weight of one acre foot of water equal to depth of water in feet.

$$\frac{\text{chemical X ppm}}{\text{LC 50 X weight of water}} = \text{depth} \qquad \text{eq. 16-1}$$

or

$$\frac{A \times B}{LC\ 50 \times C} = D$$

A—is the amount of chemical either spilled or applied to an aquatic environment in pounds in one acre foot of water.

B—is ppm of chemical.

C—is the square feet of one acre which is 43,560 times the weight of one cubic foot of water which is 6.43 pounds and this equals 2,719,450 pounds in one acre cubic foot of water.

LC 50—is the 96 hour static fish test; other hours such as 24 hour test are acceptable.

D—is the depth of water which would contain the concentration of a chemical to produce an LC 50 in fish.

Remember, the equation is calculated in one acre cu ft of water. By changing D, one can change the hours needed to obtain and LC 50. This could produce a safety factor that is acceptable to save fish or to prevent a kill.

Example:

If one pound of 1,2,4-Trichlorobenzene is spilled in water, what depth of water is required to kill the fish? The chemical has a 96 hour LC 50 of 3.36 mg/L (ppm) and a 24 hr LC 50 of 13 mg/L (ppm).

$$\frac{(1\ \text{lb}) \times 1{,}000{,}000)}{(2{,}719{,}450\ \text{lb}) \times (3.36\ \text{ppm})} = \text{D/ft}$$

D = 0.109 feet of water. This means that if one pound of this chemical is present in 0.109 ft of water, it will kill fish in 96 hours, which is the 96 hour LC 50.

$$\frac{(1\ \text{lb}) \times 1{,}000{,}000)}{(2{,}719{,}450\ \text{lb}) \times (13\ \text{ppm})} = \text{D/ft}$$

D = 0.02828 feet of water. This means that if one pound of this chemical is present in 0.02828 feet of water, it will kill fish in 24 hours, which is the 24 hr LC 50.

Knowing this information, one can try to save the aquatic organisms—in this case fish.

If the 24 hour LC 50 of 13 ppm will not provide enough time to save the fish by means other than dilution, then that may be the best way to go. Add enough water to dilute the concentration to obtain an LC 50 of 96 hours or 3.36 ppm. This will increase the safety factor (SF) to 96 hours. This has a ratio of 13 to 3.36 ppm, which is 3.86 SF over the time span.

Thus the time needed to try to save the fish was increased 3.86 fold. The D in the equation can be used in the same manner. At an LC 50 of 24 hr, the D was 0.02828 feet and at LC 50 at 96 hours was 0.109 feet, thus is the ratio equalling 3.85 SF.

If we had an unknown or wanted to increase the LC 50 of the above to 168 hours (7 days) to save the fish, it can be done. By taking the ratio of the LC 50's 168 to 96 and multiplying by the D above, we can do it.

$$\frac{(LC\ 50\ 168\ hr) \times (D\ of\ LC\ 50\ 96\ hr)}{(LC\ 50\ 96\ hr)} = D$$

D = 0.19796 feet of water needed to prevent fish kill up to 168 hours, a delay that may be used to save the fish.

The equation assumes no breakdown nor adsorption of the chemical to soil. All is present for exposure during this timeframe.

The best part of this prediction technique is that if one knows the given amount of a chemical spilled into a given amount of water, and also knows the LC 50, then perhaps the fish can be saved. If only the LC 50 and water amount is known, then use a best case scenario in the equation to save the fish. It is far better to error on the side of safety.

Rainfall may have only minor effects, unless a downpour. Anyway, let's study an equation that considers runoff.

If the annual rainfall is 6 inches (.5 ft) then the RSF can be calculated by the following equation.

$$\frac{rainfall}{D} = SF \qquad\qquad \text{eq. 16-2}$$

Using the previous 96 hr LC 50, D was found to be 0.109 ft, and at 24 hrs D = 0.02828 ft. Now use the equation.

$$\frac{0.5 \text{ ft}}{0.109 \text{ ft}} = 4.587 \text{ SF}$$

$$\frac{0.5 \text{ ft}}{0.02828 \text{ ft}} = 17.68 \text{ SF}$$

As expected, the SF for the 24 hr LC 50 with 0.5 ft of rainfall is greater than that of the 96 hr LC 50. The dilution was greater.

Now consider that the chemical was on land and the 0.5 feet of rainfall caused a 40% runoff of the chemical into the aquatic environment. Is the picture getting clearer on how this equation can be used?

The new equation is

$$\frac{\text{Rainfall}}{(D) \times (\% \text{ chemical runoff})} = \text{SF} \qquad \text{eq. 16-3}$$

Using the previous LC 50,s we now have

$$\frac{(0.5 \text{ ft})}{(0.109 \text{ ft}) \times (40\%)} = 1.8348 \text{ SF}$$

$$\frac{(0.5 \text{ ft})}{(0.02828 \text{ ft}) \times (40\%)} = 7.072 \text{ SF}$$

The SF is greater for the 24 hr than that of the 96 hr LC 50 which is expected.

Okay, what does all this mean?

If one can reduce runoff to lower the presence of contamination, then one can save the fish. The SF under different conditions is considered below. To demonstrate this, let's use only the D factor for the 96 hr LC 50 which was previously determined to be 0.109 ft.

Rainfall to aquatic environment

$$\frac{1 \text{ ft}}{0.109 \text{ ft}} = 9.174 \text{ SF} \quad \text{or} \quad \frac{3 \text{ ft}}{0.109 \text{ ft}} = 27.523 \text{ SF}$$

The 27.523 SF is greater and was expected with the greater amount of rainfall.

Runoff of chemical from land

low rain & runoff   high rain & low runoff

$$\frac{1 \text{ ft}}{(0.109 \text{ ft})(0.5\%)} = 1834.8 \text{ SF} \quad \text{or} \quad \frac{3 \text{ ft}}{(0.109 \text{ ft})(0.5\%)} = 5509.5 \text{ SF}$$

low rain & runoff   high rain & low runoff

$$\frac{1 \text{ ft}}{(0.109 \text{ ft})(10\%)} = 91.74 \text{ SF} \quad \text{or} \quad \frac{3 \text{ ft}}{(0.109 \text{ ft})(10\%)} = 275.23 \text{ SF}$$

The results can be expected as the SF is greater for low runoff and low rainfall. In areas of pesticide use, the annual rainfall and expected runoff can be used to predict SF for fish and prevent fish kill. The same would apply to spill on land or in water.

Controls to reduce fish kill are:

1. Reduce the amount of chemical in the environment, and
2. Reduce slop or control runoff from land into the aquatic environment.

Factors that can be considered for pesticides are the dissipation, sorption and breakdown rates. For example, if the half life of a parent chemical is 12 days, then only 50% of a parent chemical applied could runoff at day 12. This factor can be incorporated into the equation.

A mechanism has now been presented on how to protect fish and other aquatic organisms in the event of a spill or runoff into waters.

Chapter 17
Chemical Exposure In and Out Door

Are you aware that paints contain hundreds of toxic chemicals and over a hundred potential carcinogens? We know about lead and mercury, but what about the other chemicals (i.e., organic chemicals, metals, etc.)? John Hopkins University conducted a study and found paints to contain greater than 300 toxic chemicals and about 150 chemicals that are potentially carcinogenic.[1] Chemicals in paints have caused leukemia, cancer, brain defects, psychological disorders, etc. Thus painters, artists, and homeowners are vulnerable to these toxic chemicals. Will breathing paint fumes shorten our life span? What will the aggregate, the antagonistic and the synergistic effects be, due to our exposure to paint fumes and thousands of other toxic chemicals that are allowed in our environment?

ARE YOU POISONING YOURSELF?

Have you considered what might be in your toothpastes, shampoos, baby products, bath soaps, sunblocks, or hand lotions? Did you know that these products often contain pesticides? Would you wash your face in a pesticide, in an antifreeze, or in deicing fluids used on airport runways? Of course not. But you may have allowed yourself to be exposed to similar chemicals. Propylene glycol is a pesticide, an antifreeze, a deicer, etc. It is also an ingredient used in some moist towelettes, body lotions, etc. Below is a partial list of some of the pesticides that have been found on a few product labels.

CHEMICALS	*HL	S	SB	TP	SP	D	MW	WT
Ammonium lauryl sulfate	B	A			A			
Benzyl alcohol			A		B			
BHT	A	A				A		
Boric acid					A			
Calcium peroxide				A				
Cetyl alcohol	A,B							
Citric acid		A,B			A,B			
EDTA					A			
EDTA—Disodium EDTA	A		A		A	A		
EDTA—Tetrasodium EDTA		A,B			A			
Glycerin glycol					A			
Hydrogenated caster oil				A				
Hydrogenated vegetable oil?			A					
Methylparaben	A,B	A,B	A	A	A			A
Petrolatum	A							
Propylene glycol	A,B				A	A		A
Propylparaben	A,B	A	A		A			
Salicylic acid		A						
Sodium dodecyl-benzylsulfonate		A						
Sodium lauryl sulfate		A		A				
Sodium monofluorophosphate				A				
Sodium perborate				A				
Sodium phosphate					A			
Sorbic acid							A	
Triethanolamine	A	A	A					

* The code for the letter(s) used in the chart are baby shampoo (BS), hand lotion (HL), soap (S), sunblock (SB), toothpaste (TP), shampoo (SP), deodorant (D), wash towelette (WT), mouth wash (MW), adult (A) and baby (B).

The listing of these pesticides does not infer that you will have a toxic reaction from them. Pesticides are poisons and are regulated by the Environmental Protection Agency (EPA) under the Federal Insecticide, Fungicide, and Rodenticide Act (FIFRA). Products containing these pesticides do not make any claims as to pesticidal action, therefore, they are not regulated under FIFRA or by the EPA. This partial list is given for consumer awareness, so that you can be a better informed consumer.

There are also coloring chemicals used in some of these products which may be hazardous; however, only time will tell how hazardous they can be to us.

Exposure to some chemicals may affect how we think, feel, behave and our immune systems, as well as cause allergic reactions, etc. Most of these effects may never be identified because they are so subtle. The aggregate, antagonistic, and synergistic, and cometabolism and cobiometabolism effects, that may be caused by our exposure to these and other chemicals are not considered.

ARE WE POISONING OURSELVES?

In March 1990, Dr. Russell Jaffe and Marshall Hoffman of the Serammune Physicians Lab extrapolated from a base of 8,000 subjects that 16 million Americans will show impaired immune disorders due to the effects of pesticides they breathe. The pesticides include the carbamates, organophosphates, and halogenated classes [1].

They predicted that 5 million Americans are at risk of getting runny eyes and itchy skin; in some cases shock and even death may occur. They also predicted that 11 million Americans will react with muscle and joint pain, hives, etc., and that 500,000 Americans will get asthma, bronchitis, eczema and/or migraines.

Increasing susceptibility to the body's immune system from pesticides and other chemicals can reduce the body's ability to properly repair itself. Chronic impaired immune systems could increase our risk of heart disease and cancer. The aggregate effects, the antagonistic effects, and the synergistic effects are not even considered. These effects on humans and other animals are not known. Dr. Jaffe's report did not identify which pesticides had what effects nor what the additive effects may be.

To further support Dr. Jaffe's and Mr. Hoffman's findings, the EPA conducted a pesticide exposure study called *Nonoccupational Pesticide Exposure Study* (NOPES) [2]. The EPA reports that 32 different pesticides and/or their breakdown products were found in the air of homes. These pesticides included carbamates, organophosphates, and halogenated classes. The EPA did not study human sensitivity, only the presence of pesticides in the air of residences.

Supporting the information on exposure of chemicals and the potential immune impairment, as reported by Dr. Jaffe, is a report by the EPA stating that about 300 chemicals are continuously released into the environment. These chemicals include carbamates, organophosphates, and halogenated classes and are reported by the EPA's Office of Pesticides and Toxic Substances under The Emergency Planning and Community Right-to-Know Act.

In 1979, reported in EPA's Fate of Priority Pollutants in Publicly Owned Treatment Works-Pilot Study, it was stated that about 50 chemicals are being discharged into the environment from treatment works. However, you will not see reports on the aggregate, antagonistic, or synergistic effects that chemicals have on humans and other animals.

The American Cancer Society has estimated that the United States will have 1,040,000 new cancer victims and 510,000 deaths in 1990 [3].

Many of these cancers could be caused by the chemicals we breathe, eat, or drink. Many of these cancers may be caused by chemicals as reported by Dr. Jaffe and Mr. Hoffman and by the EPA.

Another group of chemicals being released into the environment for our exposure are those from hazardous waste disposal sites. These chemicals include carbamates, organophosphates, and halogenated classes.

REFERENCES

[1] Jaffe, Russell, MD and Marshall Hoffman, "16 Million Americans Are Sensitive to Pesticides," *Health Studies Collegium*, Report 90100, March 1990.
[2] Environmental Protection Agency, *Nonoccupational Pesticide Exposure Study* (NOPES), EPA/600/3-90/003, January 1990.
[3] American Cancer Society, *Cancer Facts and Figures*, 1990.

REALTY

What toxic chemicals are we exposed to around our homes, apartments, and indoor work places? Would you purchase property if you were not informed of potential hazardous chemicals to which you could be exposed? Would you like to be informed of the possibility of toxic chemicals were you live or work? Would you like to be able to predict your potential for exposure to toxic chemicals?

I will discuss a few of the hundreds of toxic chemicals to which we could be exposed in and around property, with an emphasis on residential property. Because these chemicals are in and around our homes, I have to repeat, scientists do not study the aggregate, antagonistic, or synergistic effects that toxic chemicals have on humans and the environment. With this in mind, remember we are exposed to hundreds of toxic chemicals, not just ONE.

The EPA conducted a study, although incomplete, revealing toxic pesticides that are in the air we breathe in our homes. Our homes are contaminated with many toxic pesticides. The EPA only looked for 32 pesticides in our air and found all 32. There could be hundreds that the EPA did not look for. The study is titled *Nonoccupational Pesticide Exposure Study* (NOPES) [1]. Some of the pesticides reported have been banned or restricted from use for years,, while others are still available for home use inside and out. The pesticides that were reported are [1]:

- ortho-Phenylphenol, a lysol ingredient that, like phenols, could be toxic.
- Captan (N-Trichloromethylthio-4-cyclohexane-1,2-dicarboximide) also N-[(trichloromethyl) thio] tetrahydrophthlimide (a chemical similar in structure to thalidomide which caused birth defects. Captan is a potential teratogen, and can effect the central nervous system, as well as cause other side effects. What effects will the presence of this chemical have on future generations?
- There are also other similar pesticides registered: Folpet [N-(trichloromethylthio) phthalimide], Imidan (0,0-Dimethyl-S-phthalimidomethyl phosphorodithioate) and Tetramethrin (also called phthalthrin) are all very close to thalidomide.
- Hexachlorobenzene, aldrin, alpha-BHC, carbaryl, chlordane, 2,4-D (methyl and butoxyethyl esters), 4,4'-DDD, 4,4'-DDE, 4,4'-DDT, Dicofol (DDT in another wrapper), dieldrin, gamma-BHC, methoxychlor (DDT in another wrapper), heptachlor, heptachlor epoxide, and oxychlordane are all considered hazardous because of their carcinogenic effects, and other side effects on humans and the environment. These chemicals are found in the air of homes, and in the air that we continuously breath. Most have been detected in—you guessed it—MOTHER'S MILK. Babies be aware that milk may be toxic to your health.
- Atrazine, chlorothalonil (Daconil), dacthal (DCPA), bendicarb, chlorpyrifos (Dursban and Lorsban), diazinon, dichlorvos (Vapona and DDVP), malathion, cis and trans-permethrin, ronnel, propoxur (Baygon), and resmethrin. Some of these chemicals are highly toxic and have caused death.

Discussed below are a few of the hundreds of toxic pesticides to which we could be exposed.

- 1,2-Dibromo-3-chloropropane (DBCP), a highly toxic pesticide that has been widely used since the late 1950s. This chemical has caused sterility, birth defects, and cancer. DBCP has been found in well water (ground water), human tissue, and perhaps mother's milk. The EPA cancelled DBCP in all states but one, Hawaii.
- Researchers at the University of Indiana in October of 1984 produced evidence that dioxin was present in rain. Dioxins are a mixture of 75 compounds with 2,3,7,8-tetrachlorodibenzo-p-dioxin

(tetrachlordioxin (TCDD)) being the most toxic of all dioxins in this class. The dioxin residues in rain apparently got there from incinerators.

- Data indicates that dioxin produces sterility in women, as well as having carcinogenic effects, teratogenic effects, and other effects. There is evidence that data on the hazards of dioxin were known in the early 1960s. There is a strong possibility that dioxin was being used as an experimental chemical for chemical warfare. Dioxin has been found in many pesticides, such as 2,4-D, 2,4,5-T, silvex, agent orange, pentachlorophenol (PCP), trichlorophenol (TCP), hexachlorophene, and others.
- Leptophos (0-(4-Bromo-2, 5-dichlorophenyl) O-methyl phenylphosphonothioate), a neurotoxin, causes nerve damage and damage to the central nervous system.
- Lindane (Gamma isomer of benzene hexachloride), also lindane BHC, was another pesticide widely used as an insect and moth control for homes, in dog dips, floor waxes, shelf paper, plus many more previously registered uses. Lindane residues are persistent in the environment and can be found in human tissue and mother's milk. Another chemical, benzene hexachloride (BHC), similar to lindane (BHC), also produces cancer and fetus defects.
- Mirex—Dodecachlorooctahydro-1,3,4-metheno-1H-cyclobuta [cd] pentalene—has widely been used in the southern United States to control fire ants. Mirex is a potential cancer causer and also causes birth defects. Mirex breaks down to kepone—Decachlorooctahydro-1, 3, 4-metheno-2H-cyclobuta [cd] pentalen-2-one—in the environment. Remember the presence of kepone in Virginia's James River? Kepone causes neurological effects on humans. When the use of mirex was banned, another pesticide, ferriamicide, was accepted for use to control fire ants. Ferriamicide is made from mirex; that is, it is mirex in another form. Ferriamicide breaks down to photomirex which is 100 times more toxic than mirex. These chemicals can produce cancer, liver damage, and reproductive problems. Residues of these chemicals have been detected in human tissues and it would not be surprising if the residues are found in mother's milk.
- Polychlorinated biphenyls, a group of chemicals widely referred to as PCBs, are also known as Aroclors. Aroclor, once a registered pesticide, has also been used in electrical equipment for years. PCBs have a long history of environmental contaminations. PCB products

have been contaminated with highly toxic dioxins, such as 2,3,7,8-tetrachlorodibenzo-*p*-dioxin (TCDD) and certain (dibenzofurans). For over 50 years, it was known that PCBs were carcinogens and mutagens. In addition, certain impurities often in PCBs (dioxins) can cause chloracne. Residues of PCBs have been found in air, water, fish, ground water (wells), human tissue, sperm, mother's milk, etc. To a large extent, most of the environmental catastrophes and human exposure could, and should, have been prevented. 125 million people—or about half the American population—have been exposed to PCBs.

- Toxaphene (Technical chlorinated camphene (67%-69% chlorine) a widely used pesticide, is a known carcinogen, mutagen, and teratogen. Residues of toxaphene have been found in air, water, soil, plants and animals, and in milk.

Other products that are in use in many homes contain toxic ingredients. Some of the toxic products that we bring into our homes include cleaners, detergents, polishes, bleach, ammonia, nail polish, iodine, paints, lacquer, motor oil, thinners, antifreeze, turpentine, and paper corrective fluid. Some of these products contain trichloroethylene (TCE), a known cancer-causer.

Radon, which can cause lung cancer, may be found in our air, well water, or from leaching in granite stone homes.

Asbestos (especially amosite, brown asbestos), which is used in older dwellings, can cause lung cancer. Asbestos may also be naturally occurring in soil in real and residential property.

Lead, which has been found in paint and tap water, has been reported to cause lead poisoning in children.

Sulfates discharged into the air from nearby industries have caused deaths, breathing problems, heart and lung disease symptoms, etc.

All around us are sulfur dioxide and nitrogen dioxide emissions from car exhaust, from burning firewood, and nearby industrial plants. Inhalation of these can cause respiratory problems. These chemicals react with moisture to form sulfuric acid and nitric acid, more commonly known as ACID RAIN.

Formaldehyde fumes when inhaled can cause cancer. Formaldehyde was used as insulation in the walls of homes, in carpets, and in other home fabrics.

Vinyl chloride fumes, if inhaled, can cause cancer and are contained in many of our plastic products.

Dioxins, one of the moist toxic chemicals invented by humans, has been detected in our paper products (paper plates, cups, milk cartons, etc.), and some scientists have said it is a by-product of firewood combustion.

Toxic ash in air, discharged from nearby incinerators containing lead, mercury, cadmium, etc., is available for our inhalation.

Chemicals from water treatment plants may release ammonia and chlorine gas, which are toxic if inhaled. Ammonia can react with chlorine to produce an explosion, and chlorine gas can also react with water to produce hydrochloric acid.

Well water may be contaminated with pesticides, heavy metals, oily products, and industrial waste.

Chemicals from hazardous waste disposal and storage sites have contaminated our air and well water.

Western mining states have heavy metals in their air (with dust), in their well water, in their soils, in some construction materials that used the mine tailings, and in ground and surface water.

Some products may emit electromagnetic fields (EMF) which could cause cancer. These products include computer screens, televisions, clocks, and high tension electrical wires.

Dr. Jaffe and Mr. Hoffman of the Serammune Physicians Lab extrapolated that 16 million Americans will show impaired immune disorders due to the effects of pesticides they inhale [2]. They did not consider , nor do I believe that they could consider, the aggregate, antagonistic, and synergistic effects of some or all the chemicals to which we could be exposed.

EPA has stated that there are over 300 hundred chemicals released into the environment [3]. This does not include all of the pesticides which are intentionally applied in the environment.

REFERENCES

[1] Environmental Protection Agency, *Nonoccupational Pesticide Exposure Study* (NOPES), EPA/600/3-90/003, January 1990.
[2] Jaffe, Russell, MD and Marshall Hoffman, "16 Million Americans Are Sensitive to Pesticides," *Health Studies Collegium*, Report 90100, March 1990.
[3] Environmental Protection Agency, *The Emergency Planning and Community Right-to-Know Act*, EPA/560/4-90-002, December 1989.

WATER

If current contaminated waters are not cleansed, the old saying "water water everywhere and not a drop to drink" will come true. Since we are what we eat, drink, and inhale, we will become contaminated and this contamination will be passed to future generations.

Lake Michigan is contaminated with pesticides and industrial pollutants—no swimming is allowed in some areas and no eating of fish is advised for most of the lake.

Some of Florida's waters contain such high levels of mercury that the contaminated fish cannot be eaten. Diazinon is also being discharged into the ocean.

New York has the famous Love Canal contamination by Hooker Chemical.

Long Island, Wisconsin, and Florida have one of the most acutely toxic pesticides in their ground water, called aldicarb (Temik). When this chemical was first introduced to the market, several pesticide applicators were poisoned and hospitalized.

Tennessee has cancer-causing chemicals in some of the waters.

The western mining states are having their waters polluted with heavy metals.

The San Joaquin Valley and its ground waters are contaminated with pesticides. Adverse effects have been reported in children exposed to the pesticides.

California, Connecticut, Florida, Hawaii, Massachusetts, South Carolina, and Texas have had their ground water contaminated with the pesticide ethylene dibromide (EDB).

- DBCP has been found in ground water in San Joaquin Valley.
- Baltimore, Maryland has a chromium plant polluting its water.
- Hundreds of petroleum spills are contaminating waters across the country.
- Chemicals from junk yards leach into ground water and run off into streams.
- Radioactive chemicals and nuclear waste contaminate our waters.
- Acid rain will have a tremendous long term effect on our waters and forests.
- In 1976, the highly creditable oceanographer Jacques Cousteau said that there was a danger that oceans would be "dead before another 50 years have passed" [1].He may be proven correct if we continue

polluting the oceans at our current rate. Residues of polychlorinated biphenyls (PCBs) have been found in dolphins at extremely high levels. Recently, Florida reported that diazinon was being discharged in the Atlantic Ocean, and was running off of lawns into the ocean. What will happen to us when we kill the waters and aquatic life? Are we running out of time or has the time of no return already been reached?

REFERENCES

[1]   Armstrong, Garner Ted, "The Death of the Oceans," *Plain Truth,* January 1976.

With a little information, the use of this book will enable you to make predictions on what will happen to chemicals in the environment.

Chapter 18
Predictive Equations for
Organic Chemical Compounds

Chapters previous to this one have discussed using predictive techniques for leaching, octanol water partition coefficient, water solubility, soil sorption, and bioconcentration and bioaccumulation. Mathematical equations were also given for predicting the fate and transport of pesticides. The equations now presented are to predict fate and transport for chemicals other than pesticides, although pesticides will fit into some of the following equations. Again, these equations are for organic chemical compounds.

Several equations may be used for more than one class of organic chemical. You may use one or more equations that fit the situation. I will not repeat information from previous chapters. You may refer to these chapters, however, for explanations.

Refer to Table 4-2, Predicting Where the Chemical Goes, to make your predictions. When applicable, also use Chapter 14 techniques in your predictions.

The following equations [1] are now given.

$$\log S = -1.37 \log K_{ow} + 7.26 \qquad \text{eq. 18.1}$$
Used for aromatics and chlorinated hydrocarbons & unit is in mol/L

$$\log 1/S = 1.113 \log K_{ow} - 0.926 \qquad \text{eq. 18.2}$$
Used for alcohols & unit is in mol/L

$$\log 1/S = 1.229 \log K_{ow} - 0.72 \qquad \text{eq. 18.3}$$
Used for ketones & unit is in mol/L

$$\log 1/S = 1.013 \log K_{ow} - 0.52 \qquad \text{eq. 18.4}$$
Used for esters & unit is in mol/L

$$\log 1/S = 1.182 \log K_{ow} - 0.935 \qquad \text{eq. 18.5}$$
Used for ethers & unit is in mol/L

$$\log 1/S = 1.221 \log Kow - 0.832 \qquad \text{eq. 18.6}$$
Used for alkyl halides & unit is in mol/L

$$\log 1/S = 1.294 \log Kow - 1.043 \qquad \text{eq. 18.7}$$
Used for alkynes & unit is in mol/L

$$\log 1/S = 1.294 \log Kow - 0.248 \qquad \text{eq. 18.8}$$
Used for alkenes & unit is in mol/L

$$\log 1/S = 1.237 \log Kow + 0.248 \qquad \text{eq. 18.9}$$
Used for alkanes & unit is in mol/L

$$\log 1/S = 0.996 \log Kow - 0.339 \qquad \text{eq. 18.10}$$
Used for aromatics & unit is in mol/L

$$\log 1/S = 1.214 \log Kow - 0.85 \qquad \text{eq. 18.11}$$
Used for propionitriles & unit is in mol/L

$$\log 1/S = 1.339 \log Kow - 0.978 \qquad \text{eq. 18.12}$$
Used for propionitriles & unit is in mole/L

$$\log 1/S = -2.38 \log Kow + 12.9 \qquad \text{eq. 18.13}$$
Used for phosphate esters & unit is in μ mol/L

$$\log S = -0.962 \log Kow + 6.50 \qquad \text{eq. 18.14}$$
Used for halogenated hydrocarbons & units in μ mol/L

$$\log Koc = -0.54 \log S + 0.44 \qquad \text{eq. 18.15}$$
Used for aromatics & polynuclear aromatics
(S or WS has to be in mole fraction)

$$\log Koc = -0.557 \log S + 4.277 \qquad \text{eq. 18.16}$$
Used for chlorinated hydrocarbons
(S or WS has to be in μ mol/L to calculate)

$$\log Koc = 0.937 \log Kow - 0.006 \qquad \text{eq. 18.17}$$
Used for aromatics, PNAs, triazines & dinitroaniline
herbicides

$$\log Koc = 0.524 \log Kow + 0.855 \qquad \text{eq. 18.18}$$
Used for substituted phenylureas &
alkyl-N-phenylcarbamates

$$\log BCF = 0.76 \log Kow - 0.23 \qquad \text{eq. 18.19}$$
Used for a wide variety of organic chemicals

$$\log Koc = 0.937 \log Kow - 0.006 \qquad \text{eq. 18.20}$$
Used for aromatics, triazines, dinitroanilines & PNAs

$$\log Koc = 1.00 \log Kow - 0.21 \qquad\qquad \text{eq. } 18.21$$
Used for aromatics & PNAs

$$\log Koc = 0.524 \log Kow + 0.855 \qquad\qquad \text{eq. } 18.22$$
Used for phenylureas & alkyl-N-phenycarbamates

EXERCISE

For each of the following use the data given and calculate additional data to make a prediction on fate and transport. You may give an in-depth evaluation or a short abbreviated one, depending on your needs. I have elected to present short and different types to acquaint the reader with different ideas.

Unless otherwise mentioned, all the data given below was found in the *EPA Handbook* [2]. Mentioning of the terms "suspected carcinogen," "questionable carcinogen," "confirmed human carcinogen," or "confirmed carcinogen" were found in the book *Hazardous Chemicals Desk Reference* [3].

#1. Acenaphthene: mol. wt. 154.21 g/mol, log Kow 3.92, WS 3.42 X 10^{00} mg/L at 25° C, & VP 1.0 X 10^{01} mm Hg at 131° C.

#2. Acetone: mol. wt. 58.08 g/mol, log Kow –0.24, WS miscible, & VP 2.66 X 10^{-02} mm Hg at 25° C.

#3. Acetonitrile: mol. wt. 41.05 g/mol, log Kow –0.34, WS miscible, & VP 7.4 X 10^{-01} mm Hg at 20° C.

#4. Acrolein: mol. wt. 56.06 g/mol, log Kow –0.09, WS 2.08 X 10^{05} mg/L at 20° C, & VP 2.10 X 10^{02} mm Hg at 20° C.

#5. Aminobiphenyl or 1,1'-biphenyl-4-amine: mol. wt. 169.23 g/mol, log Kow 2.78, WS slightly soluble, & VP 6 X 10^{05} mm Hg at 25° C to 30° C.

#6. 1,2-Benzenedicarboxlic acid, dioctyl ester: mol. wt. 390.57 g/mol, log Kow 9.2, WS 3 X 10^{00} mg/L at 25° C, & VP <2 X 10^{-01} mm Hg at 150° C.

#7. Benzenemethanol or benzene alcohol: mol. wt. 108.14 g/mol, log Kow 1.10, WS 3.5 X 10^{04} mg/L at 20° C, & VP 1.5 X 10^{01} mm Hg at 25° C.

#8. Benzo(k)fluoranthene: mol. wt. 252.32 g/mol, log Kow 6.84, WS 5.5 X 10^{-04} mg/L at 25° C, & VP 9.59 X 10^{-11} mm Hg at 20° C.

#9. Benzo(ghi)perylene: mol. wt. 276.34 g/mol, log Kow 7.23, WS 2.6 X 10^{-04} mg/L at 25° C, & VP 1 X 10^{-10} mm Hg at 20° C.

#10. Bromobenzene: mol. wt. 157.01 g/mol, log Kow 2.99, WS 4.5 X 10^{02} mg/L at 30°C, & VP 3.3 X 10^{00} mm Hg at 30°C.

#11. Butylbenzene: mol. wt. 134.22 g/mol, log Kow 4.2, WS 1.4 X 10^{01} mg/L (UT),& VP 1 X 10^{00} mm Hg at 23°C.

#12. Chlorobenzilate or ethyl 4,4'-dichlorobenzilate; this is an ethyl ester: mol. wt. 325.20 g/mol, log Kow NA, WS 1.0 X 10^{01} mg/L at 20° C, & VP 22 X 10^{-06} mm Hg at 20° C.

#13. 4-Chlorobenzamine: mol. wt. 127.57 g/mol, log Kow 1.83, WS 3.9 X 10^{03} mg/L at 25° C, & VP 5.91 X 10^{-02} mm Hg at 25° C.

#14. 2-Chloronaphthalene: mol. wt. 162.62 g/mol, log Kow 4.12, WS 6.74 X 10^{00} mg/L at 25° C, & VP 1.7 X 10^{-02} mm Hg at 20° C.

#15. 4-Chlorophenyl phenyl ether: mol. wt. 204.66 g/mol, log Kow 4.08, WS 3.3 X 10^{00} mg/L at 25° C, & VP 2.7 X 10^{-03} mm Hg at 25° C.

#16. 2-Chlorotoluene or 1-chloro-2-methylbenzene: mol. wt. 126.59 g/mol, log Kow 3.42, WS slightly, & VP 2.7 X 10^{-00} mm Hg at 20° C.

#17. 1,1-Dichloroethane: mol. wt. 98.96 g/mol, log Kow 1.79, WS 5.5 X 10^{03} at 20° C, & VP 2.34 X 10^{02} mm Hg at 25° C.

#18. Dichloroethylether or 1,1'-oxybis (2-chloroethane): mol. wt. 143.01 g/mol, log Kow 1.58, WS 1.02 X 10^{04} mg/L [UT], & VP 7.14 X 10^{-01} mm Hg at 20° C.

#19. 1,4-Dioxane: mol. wt. 88.11 g/mol, log Kow –.42, WS miscible, & VP 3.71 X 10^{01} mm Hg at 25° C.

#20. Diphenylamine: mol. wt. 169.23 g/mol, log Kow 3.22, & WS 3 X 10^{02} mg/L at 25° C.

#21. Endosulfan I: mol. wt. 406.95 g/mol, log Kow 3.55, WS 3.2 X 10^{-01} mg/L at 22° C, & VP 1 X 10^{-06} mm Hg at 25° C.

#22. 2-Hexanone or methyl butyl ketone: mol. wt. 100.16 g/mol, log Kow 1.38, WS 3.5 X 10^{04} mg/L at 20° C, & VP 2 X 10^{00} mm Hg at 20° C.

#23. Isobutyl alcohol or 2-methyl-1-propanol: mol. wt. 74.12 g/mol, log Kow 0.83, WS 9.5 X 10^{04} mg/L at 18° C, & VP 1 X 10^{01} mm Hg at 25° C.

#24. Lindane: mol. wt. 290.83 g/mol, log Kow 3.24, WS 7.0 X 10^{00} mg/L at 20° C, & VP 9.4 X 10^{-06} mm Hg at 20° C.

#25. 2-Naphthylamine: mol. wt. 143.19 g/mol, log Kow 2.25, WS 5.86 X 10^{02} mg/L at 20 to 30° C, & VP 1 X 10^{-00} mm Hg at 108° C.

#26. Tetraethyl dithiopyrophosphate or tetraethyl ester of thiodiphosphoric acid: mol. wt. 322.31 g/mol, log Kow 3.83, WS 2.5 X 10^{01} mg/L [UT], & VP 1.7 X 10^{-04} mm Hg at 20° C.

#27. Vinylidene chloride: mol. wt. 96.94 g/mol, log Kow 1.48, WS 2.1 X 10^{02} mg/L at 25° C, & VP 5.91 X 10^{02} mm Hg at 25° C.

ANSWERS

#1.   Acenaphthene: mol. wt. 154.21 g/mol, log Kow 3.92, WS 3.42 X 10^{00} mg/L at 25° C, & VP 1.0 X 10^{01} mm Hg at 131° C.

Predictive Equation Analysis

$$\log BCF = 0.76 \log Kow - 0.23 \qquad (eq.18-19)$$
$$= 0.76 \log (Kow\ 3.92) - 0.23$$
$$= 2.7492$$
$$BCF = 561.31$$

$$\log Koc = 0.937 \log Kow - 0.006 \qquad (eq.\ 18.20)$$
$$= 0.937 (\log Kow\ 3.92) - 0.006$$
$$= 3.667$$
$$Koc = 4{,}645.58$$
$$Kow = 8{,}317.6377$$

Fate and Transport Analysis

- WS indicates that acenaphthene should adsorb to soil and bioaccumulate.
- Koc indicates that the chemical should absorbed and accumulated in the soil.
- BCF and Kow indicate that it should be lipo soluble and should bioaccumulate.
- VP indicates that it may be volatile and phototransformation may occur.

Exposure Analysis

This chemical may be persistent in the environment and, if released in the environment, food chain contamination is likely. It is lipo soluble and could bioaccumulate in the food chain. This chemical appears to be volatile and inhalation is likely.

#2.   Acetone: mol. wt. 58.08 g/mol, log Kow –0.24, WS miscible, & VP 2.66 X 10^{-02} mm Hg at 25° C.

Predictive Equation Analysis

log 1/S   = 1.229 log Kow –0.72   (eq. 18.3)
            = 1.229 (log Kow –.24 mol/L) –0.72
            = –1.015 mol/L
S   = 10.4537 mol/L
S   = 607,150.324 mg/L

log BCF = 0.76 log Kow – 0.23   (eq. 18.19)
            = 0.76 (log Kow –.24) –0.23
            = –0.412
BCF   = 0.3869
Kow   = 0.575

Fate and Transport Analysis

• WS indicates that acetone could leach, runoff, and be biodegraded.
• Kow indicates that acetone is not lipo soluble and should not bioaccumulate. This is supported by the BCF value.
• This chemical may be volatile and phototransformation may occur.

Exposure Analysis

Acetone could be soluble enough to leach, runoff, and be biodegraded. However, its volatility may preempt this by volatilizing into the environment. Since it may be volatile, exposure by inhalation is expected.

#3.   Acetonitrile: mol. wt. 41.05 g/mol, log Kow –0.34, WS miscible, & VP 7.4 X 10^{-01} mm Hg at 20° C.

Predictive Equation Analysis

log BCF = 0.76 log Kow –0.23   (eq. 18.19)
            = 0.76 (log Kow –0.34) –0.23
            = –0.4884
BCF   = 0.325
Kow   = 0.457

Fate and Transport Analysis

- Solubility indicates that acetonitrile could leach, runoff, and be biodegraded.
- BCF indicates that this chemical should not bioaccumulate and this is supported by the Kow value.
- This chemical is volatile and phototransformation may occur.

Exposure Analysis

Acetonitrile could leach, runoff, and be biodegraded in the environment. Based on its WS, BCF, and Kow, it should not bioaccumulate in the food chain. It may be volatile and exposure by inhalation could be expected.

#4. Acrolein: mol. wt. 56.06 g/mol, log Kow –0.09, WS 2.08 X 10^{05} mg/L at 20° C, & VP 2.10 X 10^{02} mm Hg at 20° C. This chemical is a questionable carcinogen and one of its expected metabolites, glycidaldehyde, is a carcinogen.

Predictive Equation Analysis

$$\log BCF = 0.76 \log Kow - 0.23 \qquad \text{(eq. 18.19)}$$
$$= 0.76 (\log Kow - 0.09) - 0.23$$
$$= -0.2948$$

BCF = 0.503

Kow = 0.813

Fate and Transport Analysis

- WS indicates that acrolein could leach, runoff, and be biodegraded.
- BCF and Kow indicate that this chemical should not bioaccumulate.
- This chemical is volatile and phototransformation may occur.

Exposure Analysis

This chemical could leach, runoff, and be biodegraded. Bioaccumulation is not expected. Based on its volatility, inhalation could be a problem. Another name for this chemical is acraldehyde. Based on its chemical structure, an aldehyde, it is predicted that this chemical could be a carcinogen. This chemical may be metabolized or degraded to glycidaldehyde, a carcinogen. This chemical is, in fact, a questionable carcinogen, while its metabolite is a carcinogen and care should be taken to prevent contamination and inhalation.

#5. 4-Aminobiphenyl or 1,1'-biphenyl-4-amine: mol. wt. 169.23 g/mol, log Kow 2.78, WS slightly soluble, & VP 6 X 10^{05} mm Hg at 25° C to 30° C. This chemical is a confirmed human carcinogen.

Predictive Equation Analysis

$$
\begin{aligned}
\log \text{BCF} &= 0.76 \log \text{ Kow } -0.23 \qquad\qquad \text{(eq. 18.19)}\\
&= 0.76 (\log \text{ Kow } 2.78) -0.23\\
&= 1.86
\end{aligned}
$$

BCF = 72.44

Kow = 602.56

Fate and Transport Analysis

- This chemical is slightly soluble and leaching and runoff is expected. Biodegradation could occur.
- Kow and BCF indicate that it should not bioaccumulate.
- Volatility could occur and phototransformation may occur.

Exposure Analysis

Based on the solubility (it is slightly soluble), this chemical could leach and runoff. The Kow and BCF indicate that this should not bioaccumulate. This chemical is volatile and inhalation is expected. Based on the chemical structure, this chemical contains amine, amino and/or biphenyl, which means it could be predicted to be a carcinogen. In fact, it is a confirmed human carcinogen.

#6. 1,2-Benzenedicarboxlic acid, dioctyl ester or di-n-octyl phthalate: mol. wt. 390.57 g/mol, log Kow 9.2, WS 3 X 10^{00} mg/L at 25° C, & VP <2 X 10^{-01} mm Hg at 150° C.

Predictive Equation Analysis

$$
\begin{aligned}
\log 1/S &= 1.013 \log \text{ Kow } -0.52 \text{ (mol/L)} \qquad \text{(eq. 18.4)}\\
&= 1.013 (\log \text{ Kow } 9.2) -0.52\\
&= 8.7996
\end{aligned}
$$

S = 6.587 X 10^{-7} mol/L

S = 6.198 X 10^{-3} mg/L

$$\log \text{BCF} = 0.76 \log \text{Kow} - 0.23 \qquad \text{(eq. 18.19)}$$
$$= 0.76 (\log \text{Kow } 9.2) - 0.23$$
$$= 6.76$$

BCF $= 5,780,900.47$

Kow $= 1,584,893,192$

Fate and Transport Analysis

- WS indicates that this chemical should not leach and be adsorbed in the soil's organic matter, and could runoff with soil particles.
- BCF and Kow indicate that bioaccumulation should occur.
- VP indicates that volatility could occur and phototransformation may occur.

Exposure Analysis

If this chemical reaches the aquatic environment, bioaccumulation is likely to occur. If it reaches any animals, bioaccumulation is likely because it is lipo soluble. Inhalation is likely because it may be volatile. A similar chemical, di-sec-octylphthalate, is a confirmed human carcinogen. Based on its chemical structure, this chemical is predicted to be a carcinogen.

#7.   Benzenemethanol or benzene alcohol: mol. wt. 108.14 g/mol, log Kow 1.10, WS 3.5 X 10^{04} mg/L at 20° C, & VP 1.5 X 10^{01} mm Hg at 25° C.

Predictive Equation Analysis

$$\log \text{BCF} = 0.76 \log \text{Kow} - 0.23 \qquad \text{(eq. 18.19)}$$
$$= 0.76 (\log \text{Kow } 1.10) - 0.23$$
$$= 1.86$$

BCF $= 72.44$

Kow $= 12.589$

Fate and Transport Analysis

- WS indicates that this chemical could leach and runoff into the aquatic environment, and it could be biodegraded.
- BCF and Kow indicate that it should not bioaccumulate in the environment.

- VP indicates that volatility could occur and phototransformation may occur.

Exposure Analysis

This chemical is water soluble and may leach and runoff. It should be biodegraded in the soil. VP indicates that exposure via inhalation may occur. Based on its chemical structure, benzene, which is a confirmed human carcinogen, may be one of its degradation or phototransformation products. Thus, it is predicted to be a carcinogen.

#8. Benzo(k)fluoranthene: mol. wt. 252.32 g/mol, log Kow 6.84, WS 5.5 X 10^{-04} mg/L at 25° C, & VP 9.59 X 10^{-11} mm Hg at 20° C. This chemical is a confirmed carcinogen.

Predictive Equation Analysis

$$
\begin{aligned}
\log S \quad &= -1.37 \log Kow + 7.26 \ (\mu \ mol/L) \qquad &\text{(eq. 18.1)} \\
&= -1.37 \ (\log Kow \ 6.84) + 7.26 \\
&= -2.1108 \\
S \quad &= 0.0077 \ \mu \ mol/L \\
S \quad &= 2.0 \ X \ 10^{-3} \ mg/L
\end{aligned}
$$

$$
\begin{aligned}
\log BCF &= 0.76 \log Kow - 0.23 \qquad &\text{(eq. 18.19)} \\
&= 0.76 \ (\log Kow \ 6.84) - 0.23 \\
&= 4.97 \\
BCF \quad &= 92{,}982.24 \\
Kow \quad &= 6{,}918{,}309.709
\end{aligned}
$$

Fate and Transport Analysis

- WS indicates that this chemical should not leach or runoff, unless with soil particles. Adsorption to soil's organic matter should occur.
- BCF and Kow indicate that bioaccumulation could occur.
- VP indicates that volatility should not occur.

Exposure Analysis

If this chemical gets in any environmental compartment it can be a problem because it could bioaccumulate. Its chemical structure indicates that it may be predicted to be a carcinogen. This chemical is, in fact, a confirmed carcinogen.

#9. Benzo(ghi)perylene: mol. wt. 276.34 g/mol, log Kow 7.23, WS 2.6 X 10^{-04} mg/L at 25° C, & VP 1 X 10^{-10} mm Hg at 20° C.

Predictive Equation Analysis

$$\text{log BCF} = 0.76 \text{ log Kow} - 0.23 \qquad \text{(eq. 18.19)}$$
$$= 0.76 \,(\text{log Kow } 7.23) - 0.23$$
$$= 5.2648$$

BCF   = 16,982,436

Kow   = 16,982,436.52

Fate and Transport Analysis

- WS indicates that this chemical should adsorb to soil's organic matter, should not leach, and may runoff with soil particles.
- BCF and Kow indicate that bioaccumulation could occur.
- VP indicates that volatility should not occur.

Exposure Analysis

Mobility in soil should be restricted by soil adsorption. This chemical should bioaccumulate in the environment. Based on its chemical structure and similar chemical structures, it is predicted to be a carcinogen.

#10. Bromobenzene: mol. wt. 157.01 g/mol, log Kow 2.99, WS 4.5 X 10^{02} mg/L at 30° C, & VP 3.3 X 10^{00} mm Hg at 30° C.

Predictive Equation Analysis

$$\text{log Koc} = 0.937 \text{ log Kow} - 0.006 \qquad \text{(eq. 18.20)}$$
$$= 0.937 \,(\text{log Kow } 2.99)$$
$$= 2.7956$$

Koc   = 624.64

$$\text{log Koc} = 1.00 \text{ log Kow} - 0.21 \qquad \text{(eq. 18.21)}$$
$$= 1.00 \,(\text{log Kow } 2.99) - 0.21$$
$$= 2.78$$

Koc   = 602.56

$$\text{log BCF} = 0.76 \text{ log Kow} - 0.23 \qquad \text{(eq. 18.19)}$$
$$= 0.76 \,(\text{log Kow } 2.99) - 0.23$$
$$= 2.04$$

BCF   = 110

Kow   = 977.237

Fate and Transport Analysis

- WS indicates that this chemical may or may not leach, runoff, or be biodegradable. It may go either way.
- Koc indicates that it should leach and runoff; thus, it should be biodegradable.
- BCF and Kow indicate that bioaccumulation may occur.
- VP indicates that volatility could occur and phototransformation may occur.

Exposure Analysis

The predictive data are contradictive but still useful in that the trend is towards bioaccumulation. Inhalation may be a problem because it may be volatile. Based on similar chemical structure, it is predicted to be a carcinogen.

#11. Butylbenzene: mol. wt. 134.22 g/mol, log Kow 4.2, WS 1.4 X 10^{01} mg/L (UT),& VP 1 X 10^{00} mm Hg at 23° C.

Predictive Equation Analysis

$$\log 1/S = 0.996 \log Kow - 0.339 \text{ (mol/L)} \qquad \text{(eq. 18.10)}$$
$$= 0.996 \text{ (log Kow 4.2)} - 0.339$$
$$= 3.8442$$
$$S = 0.0001 \text{ mol/L}$$
$$S = 13.422 \text{ mg/L}$$

$$\log BCF = 0.76 \log Kow - 0.23 \qquad \text{(eq. 18.19)}$$
$$= 0.76 \text{ (log Kow 4.2)}$$
$$= 2.962$$
$$BCF = 916$$

$$Kow = 15,848.932$$

Fate and Transport Analysis

- WS indicates that this chemical should not leach and can runoff with soil particles. It could adsorb to soil's organic matter.
- BCF and Kow indicate that bioaccumulation could occur.
- VP indicates that it may be volatile and photodegradation may occur.

Exposure Analysis

This chemical may bioaccumulate in the environment. Its WS should prevent leaching, but runoff with soil particles could occur. Inhalation may be a problem because it may be volatile. Based on similar chemical structure, it is predicted to be a carcinogen.

#12. Chlorobenzilate or ethyl 4,4'-dichlorobenzilate; this is an ethyl ester: mol. wt. 325.20 g/mol, log Kow NA, WS 1.0 X 10^{01} mg/L at 20° C, & VP 22 X 10^{-06} mm Hg at 20° C. Suspected carcinogen.

Predictive Equation Analysis

$\log S = 1.4878\ \mu\ mol/L$

$\log S\ (\mu\ mol/L) = -1.37 \log Kow + 7.26$ (eq. 18.1)
$-1.37 \log Kow = \log S - 7.26$
$1.37 \log Kow = -\log S + 7.26$
$\log Kow = \dfrac{-(\log S\ 1.4878) + 7.26}{1.37}$

$= 4.2133$
$Kow = 16,341.22$

$\log Koc = -0.557 \log S\ (\mu\ mol/L) + 4.277$ (eq. 18.16)
$= -0.557\ (\log S\ 1.4878 + 4.277$
$= 3.4483$
$Koc = 2,807.16$

$\log BCF = 0.76 \log Kow - 0.23$ (eq. 18.19)
$= 0.76\ (\log Kow\ 4.5693) - 0.23$
$= 3.2427$
$BCF = 1,748.51$

Fate and Transport Analysis

- WS indicates that this chemical should not leach, could runoff with soil particles, and could be adsorbed to soil's organic matter.
- BCF and Kow indicate bioaccumulation.
- VP indicates that volatility should not occur.

Exposure Analysis

If this chemical is released in the environment, it could bioaccumulate in the environment. Based on similar chemical structure, it is predicted to be a carcinogen. Inhalation should not be a problem, as it is should not be volatile.

#13. 4-Chlorobenzamine: mol. wt. 127.57 g/mol, log Kow 1.83, WS 3.9 X 10^{03} mg/L at 25° C, & VP 5.91 X 10^{-02} mm Hg at 25° C. This chemical is also called 4-chloroaniline and is a questionable carcinogen.

Predictive Equation Analysis

$\log S$ = $-1.37 \log$ Kow + 7.26 (μ mol/L)   (eq. 18.1)
       = -1.37 (log Kow 1.83) + 7.26
       = 4.7529
S   = 56,610.89 μ mol/L
S   = 7,217.88 mg/L

$\log$ BCF = 0.76 log Kow $-$ 0.23   (eq. 18.19)
       = 0.76 (log Kow 1.83) $-$ 0.23
       = 1,908
BCF   = 12.33
Kow   = 67.608

Fate and Transport Analysis

- WS indicates that this chemical could leach, runoff, and may be biodegraded by soil microbes.
- BCF and Kow indicate that it should not bioaccumulate.
- VP indicates that it may volatilize and phototransformation may occur.

Exposure Analysis

This chemical may reach the aquatic environment; however, bioaccumulation should not occur. Biodegradation in soil should occur. Because it may be volatile, inhalation may be a problem. This chemical is, in fact, a questionable carcinogen, and based on similar chemical structure (aniline moiety) it is also predicted to be a carcinogen.

#14. 2-Chloronaphthalene: mol. wt. 162.62 g/mol, log Kow 4.12, WS 6.74 X 10^{00} mg/L at 25° C, & VP 1.7 X 10^{-02} mm Hg at 20° C. This chemical is a confirmed human carcinogen.

Predictive Equation Analysis

log S	= −1.37 log Kow + 7.26 (μ mol/L)	(eq. 18.1)
	= −1.37 (log Kow 4.12) +7.26	
	= 1.6156	
S	= 41.2667 μ mol/L	
S	= 6.71 mg/L	
log BCF =	0.76 log Kow − 0.23	(eq. 18.19)
	= 0.76 (log Kow 4.12) −.23	
	= 2.9012	
BCF	= 791.53	
Kow	= 13,182.567	

Fate and Transport Analysis

- WS indicates that this chemical should be adsorbed to soil's organic matter, should not leach, and could runoff with soil particles.
- BCF and Kow indicate that it should bioaccumulate.
- VP indicates that it could be volatile and phototransformation may occur.

Exposure Analysis

This chemical could bioaccumulate in the environment and be adsorbed to soil's organic matter; thus, preventing it from leaching. Inhalation may be is a problem because it may be volatile. Based on similar chemical structures, it was found that many naphthalene chemicals are suspected, questionable or confirmed carcinogens; thus, it is predicted to be a carcinogen. In fact, it is a confirmed human carcinogen.

#15. 4-Chlorophenyl phenyl ether: mol. wt. 204.66 g/mol, log Kow 4.08, WS 3.3 X 10^{00} mg/L at 25° C, & VP 2.7 X 10^{-03} mm Hg at 25° C.

Predictive Equation Analysis

$$
\begin{aligned}
\log S &= -1.37 \log Kow + 7.26 \ (\mu \ mol/L) \qquad \text{(eq. 18.1)} \\
&= -1.37 \ (\log Kow \ 4.08) + 7.26 \\
&= 1.67
\end{aligned}
$$

S   = 46.82 μ mol/L

S   = 9.59 mg/L

$$
\begin{aligned}
\log BCF &= 0.76 \log Kow - 0.23 \qquad\qquad \text{(eq. 18.19)} \\
&= 0.76 (\log Kow \ 4.08) - 0.23 \\
&= 2.8708
\end{aligned}
$$

BCF   = 742.68

$$
\begin{aligned}
\log Koc &= -0.557 \log S \ (\mu \ mol/L) + 4.277 \qquad \text{(eq. 18.16)} \\
&= -0.557 \ (\log S \ 1.689) + 4.277 \\
&= 3.3362
\end{aligned}
$$

Koc   = 2,168.929

Kow   = 12,022.644

Fate and Transport Analysis

- WS indicates that this chemical should not leach, could runoff with soil particles, and could be adsorbed by soil's organic matter.
- Koc indicates that it could adsorbed in soil's organic matter.
- BCF and Kow indicate that it could bioaccumulate.
- VP indicates that volatility and phototransformation may occur.

Exposure Analysis

This chemical should not be mobile in soil. This is supported by Koc value, which indicates adsorption to soil's organic matter. If it is released in the environment it could bioaccumulate. Inhalation may be a problem, based on its volatility.

#16. 2-Chlorotoluene or 1-chloro-2-methylbenzene: mol. wt. 126.59 g/mol, log Kow 3.42, WS slightly, & VP 2.7 X 10^{-00} mm Hg at 20° C.

Predictive Equation Analysis

log S	$= -1.37$ log Kow $+ 7.26$ (μ mol/L)	(eq. 18.1)
	$= -1.37$ (log Kow 3.42) $+7.26$	
	$= 2.5746$	
S	$= 375.49\ \mu$ mol/L	
S	$= 47$ mg/L	

log BCF $=$	0.76 log Kow $- 0.23$	(eq. 18.19)
$=$	0.76 (log Kow 3.42) $- 0.23$	
$=$	2.3692	
BCF	$= 233.99$	

log Koc $=$	-0.557 log S (μ mol/L) $+ 4.277$	(eq. 18.16)
$=$	-0.557 (log S 2.575) $+ 4.277$	
$=$	2.8429	
Koc	$= 696.54$	
Kow	$= 2{,}630.268$	

Fate and Transport Analysis

- WS indicates that this chemical may go either way. It was found to be slightly soluble.
- Koc indicates that it is soluble and may leach, runoff, and be biodegraded.
- BCF and Kow indicate that is bioaccumulated.
- VP indicates that it may be volatile and phototransformation may occur.

Exposure Analysis

The data are very close and real data are needed. Perhaps it is better to error on the safe side and predict that this chemical will bioaccumulate in the environment and inhalation may be a problem.

#17. 1,1-Dichloroethane: mol. wt. 98.96 g/mol, log Kow 1.79, WS 5.5 X 10^{03} at 20° C, & VP 2.34 X 10^{02} mm Hg at 25° C. This chemical is a questionable carcinogen.

Predictive Equation Analysis

$$\log \text{BCF} = 0.76 \log \text{Kow} - 0.23 \qquad \text{(eq. 18.19)}$$
$$= 0.76 (\log \text{Kow } 1.79) - 0.23$$
$$= 1.1304$$
$$\text{BCF} = 13.5$$

$$\log \text{Koc} = -0.557 \log \text{S } (\mu \text{ mol/L}) + 4.277 \qquad \text{(eq. 18.16)}$$
$$= -0.557 (\log \text{S } 4.74) + 4.277$$
$$= -2.643$$
$$\text{Koc} = 43.054$$
$$\text{Kow} = 61.659$$

Fate and Transport Analysis

- WS indicates that this chemical could leach, runoff, and be biodegraded in soil.
- BCF and Kow indicate that it should not bioaccumulate.
- VP indicates that it may volatilize and phototransformation may occur.

Exposure Analysis

This chemical should not persist in the environment and bioaccumulation should not occur. Since it is volatile, inhalation may be a problem, as this chemical is a questionable carcinogen.

#18. Dichloroethylether or 1,1'-oxybis (2-chloroethane): mol. wt. 143.01 g/mol, log Kow 1.58, WS 1.02 X 10^{04} mg/L [UT], & VP 7.14 X 10^{-01} mm Hg at 20° C. This chemical is a questionable carcinogen.

Predictive Equation Analysis

$$\log 1/\text{S} = 1.182 \log \text{Kow} - 0.935 \text{ (mol/L)} \qquad \text{(eq. 18.5)}$$
$$= 1.182 (\log \text{Kow } 1.58) - 0.935$$
$$= 8.5617$$
$$\text{S} = 0.1168 \text{ mol/L}$$
$$\text{S} = 16,703.568 \text{ mg/L}$$

$$\log BCF = 0.76 \log Kow - 0.23 \qquad \text{(eq. 18.19)}$$
$$= 0.76 (\log Kow\ 1.58) - 0.23$$
$$= 0.97$$

BCF = 9.35

Kow = 38.019

Fate and Transport Analysis

- WS indicates that this chemical could leach, runoff, and be biodegraded in soil.
- BCF and Kow indicate it should not bioaccumulate.
- VP indicates that it may volatilize and phototransformation may occur.

Exposure Analysis

This chemical should not persist in the environment and bioaccumulation should not occur. Since it may be volatile, inhalation may be a problem, as this chemical is a questionable carcinogen.

#19. 1,4-Dioxane: mol. wt. 88.11 g/mol, log Kow –.42, WS miscible, & VP 3.71 X 10^{01} mm Hg at 25° C. This chemical is a confirmed carcinogen.

Predictive Equation Analysis

$$\log BCF = 0.76 \log Kow - 0.23 \qquad \text{(eq. 18.19)}$$
$$= 0.76 (\log Kow\ -0.42) - 0.23$$
$$= -0.5492$$

BCF = 0.282

$$\log Koc = 1.00 \log Kow - 0.21 \qquad \text{(eq. 18.21)}$$
$$= 1.00 (\log Kow\ -0.42) - 0.21$$
$$= -0.63$$

Koc = 0.2344

Kow = 0.38

Fate and Transport Analysis

- WS indicates that this chemical is miscible and it could leach, runoff, and be biodegraded in soil.
- BCF and Kow indicate that it should not bioaccumulate.

- VP indicates that it may be volatile and phototransformation may occur.

Exposure Analysis

This chemical should not persist in the environment and bioaccumulation should not occur. Since it may be volatile, inhalation may be a problem, as this chemical is a confirmed carcinogen.

#20. Diphenylamine: mol. wt. 169.23 g/mol, log Kow 3.22, & WS 3 X 10^{02} mg/L at 25° C.

Predictive Equation Analysis

$$\log BCF = 0.76 \log Kow - 0.23 \qquad \text{(eq. 18.19)}$$
$$= 0.76 (\log Kow \ 3.22) - 0.23$$
$$= 2.2172$$

BCF   = 164.892

Kow   = 1,659.587

Fate and Transport Analysis

- WS indicates that this chemical could leach, runoff, and be biodegraded.
- BCF and Kow indicate it could bioaccumulate.

Exposure Analysis

There are contradictions in the predictive data. It is an amine and a diphenyl; thus, it is predicted to be a carcinogen. We should error on the side of safety and predict that bioaccumulation may occur.

#21. Endosulfan I: mol. wt. 406.95 g/mol, log Kow 3.55, WS 3.2 X 10^{-01} mg/L at 22° C, & VP 1 X 10^{-06} mm Hg at 25° C. This chemical is a questionable carcinogen.

Predictive Equation Analysis

$$\log Koc = -0.557 \log S \ (\mu \ mol/L) + 4.277 \qquad \text{(eq. 18.16)}$$
$$= -0557 (\log S \ 0.099) + 4.277$$
$$= 4.332$$

Koc   = 21,478.30

$$\log BCF = 0.76 \log Kow - 0.23 \qquad (eq.\ 18.19)$$
$$= 0.76\ (\log Kow\ 3.55) - 0.23$$
$$= 2.468$$

BCF $= 293.77$

Kow $= 3,548.134$

Fate and Transport Analysis

- WS indicates that this chemical should not leach, but may runoff with soil particles and could be adsorbed in soil's organic matter.
- Koc indicates soil adsorption could occur.
- BCF and Kow indicate that bioaccumulation can occur.
- VP indicates that it should not volatilize.

Exposure Analysis

This chemical could be persistent in the environment and bioaccumulation may occur. It is a questionable carcinogen. Care should be taken to prevent exposure, especially via the intake of contaminated food, as it is a pesticide residue.

#22. 2-Hexanone or methyl butyl ketone: mol. wt. 100.16 g/mol, log Kow 1.38, WS 3.5 X 10^{04} mg/L at 20° C, & VP 2 X 10^{00} mm Hg at 20° C.

Predictive Equation Analysis

$$\log 1/S = 1.133 \log Kow - 0.926\ (mol/L) \qquad (eq.\ 18.3)$$
$$= 1.133\ (\log Kow\ 1.38) - 0.926$$
$$= 4.07$$

S $= 0.245$ mol/L

S $= 24,589.76$ mg/L

$$\log BCF = 0.76 \log Kow - 0.23 \qquad (eq.\ 18.19)$$
$$= 0.76\ (\log Kow\ 1.38) - 0.23$$
$$= 0.8188$$

BCF $= 6.589$

Kow $= 23.988$

Fate and Transport Analysis

- WS indicates that this chemical could leach, runoff, and be biodegraded.

- BCF and Kow indicate that it should not bioaccumulate.
- VP indicates that it may volatilize and phototransformation may occur.

Exposure Analysis

This chemical should not persist in the environment and should not bioaccumulate. Inhalation may be a problem.

#23. Isobutyl alcohol or 2-methyl-1-propanol: mol. wt. 74.12 g/mol, log Kow 0.83, WS 9.5 X 10^{04} mg/L at 18° C, & VP 1 X 10^{01} mm Hg at 25° C. This chemical is a questionable carcinogen.

Predictive Equation Analysis

$$\log 1/S = 1.133 \log \text{Kow} - 0.926 \text{ (mol/L)} \qquad \text{(eq. 18.3)}$$
$$= 1.133 (\log \text{Kow } 0.83) - 0.926$$
$$= 0.9949$$

S   = 1.005 mol/L
S   = 74,490 mg/L

$$\log \text{BCF} = 0.76 \log \text{Kow} - 0.23 \qquad \text{(eq. 18.19)}$$
$$= 0.76 (\log \text{Kow } 0.83) - 0.23$$
$$= 0.4$$

BCF   = 2.52

Kow   = 6.761

Fate and Transport Analysis

- WS indicates that this chemical could leach, runoff, and be biodegraded.
- BCF and Kow indicate that it should not bioaccumulate.
- VP indicates that it may volatilize and phototransformation may occur.

Exposure Analysis

This chemical should not persist in the environment and should not bioaccumulate. Inhalation may be a problem, as this chemical is a questionable carcinogen.

#24. Lindane: mol. wt. 290.83 g/mol, log Kow 3.24, WS 7.0 X 10^{00} mg/L at 20° C, & VP 9.4 X 10^{-06} mm Hg at 20° C. This chemical is a confirmed carcinogen.

Predictive Equation Analysis

$$\log \text{Koc} = -0.557 \log \text{S (mol/L)} + 4.277 \qquad \text{(eq. 18.16)}$$
$$= -0.557 (\log \text{S } 1.3815) + 4.277$$
$$= 3.507$$
$$\text{Koc} = 3,217.57$$

$$\log \text{BCF} = 0.76 \log \text{Kow} - 0.23 \qquad \text{(eq. 18.19)}$$
$$= 0.76 (\log \text{Kow } 3.24) - 0.23$$
$$= 2.2324$$
$$\text{BCF} = 170.76$$

$$\text{Kow} = 1,737.801$$

Fate and Transport Analysis

- WS indicates that this chemical should not leach, may runoff with soil particles, and should be adsorbed to soil's organic matter.
- BCF and Kow indicate that it should bioaccumulate.
- VP indicates that it may not be volatile; however the value is close and could go either way.

Exposure Analysis

This chemical could be persistent in the environment and may bioaccumulate. It is a confirmed carcinogen. Care should be taken to prevent exposure due to ingestion and inhalation.

#25. 2-Naphthylamine: mol. wt. 143.19 g/mol, log Kow 2.25, WS 5.86 X 10^{02} mg/L at 20 to 30° C, & VP 1 X 10^{00} mm Hg at 108° C. This chemical is a confirmed carcinogen.

Predictive Equation Analysis

$$\log \text{BCF} = 0.76 \log \text{Kow} - 0.23 \qquad \text{(eq. 18.19)}$$
$$= 0.76 (\log \text{Kow } 2.25) - 0.23$$
$$= 1.48$$
$$\text{BCF} = 30.199$$

$$\text{Kow} = 177.828$$

Fate and Transport Analysis

- WS indicates that this chemical could leach, runoff, and be biodegraded in soil.
- BCF and Kow indicate that it should not bioaccumulate.
- VP indicates that it may volatilize and phototransformation may occur.

Exposure Analysis

This chemical should not be persistent in the environment. Accumulation and bioaccumulation should not occur. Inhalation may be a problem as it may be volatile. Based on its chemical structure, an amine, it is predicted to be a carcinogen. In fact, it is a confirmed carcinogen.

#26. Tetraethyl dithiopyrophosphate or tetraethyl ester of thiodiphosphoric acid: mol. wt. 322.31 g/mol, log Kow 3.83, WS 2.5×10^{01} mg/L [UT], & VP 1.7×10^{-04} mm Hg at 20° C.

Predictive Equation Analysis

$$\log 1/S = 1.013 \log \text{Kow} - 0.52 \text{ (mol/L)} \qquad \text{(eq. 18.4)}$$
$$= 1.013 (\log \text{Kow } 3.83) - 0.52$$
$$= 2289.7602$$

| S | = 0.0004 mol/L |
| S | = 129.32 mg/L |

$$\log \text{BCF} = 0.76 \log \text{Kow} - 0.23 \qquad \text{(eq. 18.19)}$$
$$= 0.76 (\log \text{Kow } 3.83) - 0.23$$
$$= 2.681$$

| BCF | = 479.51 |
| Kow | = 6,760.83 |

Fate and Transport Analysis

- WS indicates that this chemical could go either way.
- BCF and Kow indicate that it should bioaccumulate.
- VP indicates that it may be volatile.

Exposure Analysis

Mixed predictive results; however, this chemical is an organic phosphate and care should be taken to prevent exposure. Inhalation may be a problem if volatility is established. Real data are needed.

#27. Vinylidene chloride: mol. wt. 96.94 g/mol, log Kow 1.48, WS 2.1 X 10^{02} mg/L at 25° C, & VP 5.91 X 10^{02} mm Hg at 25° C. This chemical is a suspected carcinogen.

Predictive Equation Analysis

log BCF= 0.76 log Kow − 0.23   (eq. 18.19)
      = 0.76 (log Kow 1.48) − 0.23
      = 0.894
BCF   = 7.8487

log Koc = −0.557 log S + 4.277 (μ mol/L)   (eq. 18.16)
      = −0.557 (log Kow 3.3357) + 4.277
      = 2.419
Koc   = 262

Kow   = 30.20

Fate and Transport Analysis

• WS indicates that this chemical could go either way.
• Koc indicates that it should not adsorb to soil.
• BCF and Kow indicate that it should not bioaccumulate.
• VP indicates that volatility could occur and phototransformation may occur.

Exposure Analysis

This chemical should not be persistent in the environment and should not bioaccumulate. However, inhalation may be a problem since it is, in fact, a suspected carcinogen. Based on the chemical structure, a vinyl structure, it is also predicted to be a carcinogen.

The following are a list of chemicals given for additional information.

#1. 2-Acetylaminofluorene or N-9H-fluoren-2-yl-acetamide: mol. wt. 223.27 g/mol, log Kow 3.28, & WS insoluble. This chemical is a questionable carcinogen.

Predictive Equation Analysis

log BCF= 0.76 log Kow − 0.23   (eq. 18.19)
      = 0.76 (log Kow 3.28) − 0.23
      = 2.2628
BCF   = 183.147

$$\log Koc = 0.937 \log Kow - 0.006 \qquad \text{(eq. 18.17)}$$
$$= 0.937 \, (3.28) - 0.006$$
$$= 3.0628$$

Koc $\quad = 1,167.777$

Kow $\quad = 1,905.46$

Fate and Transport Analysis

- WS (insoluble) indicates that this chemical should not leach, could runoff with soil particles, and should adsorb to soil's organic matter.
- Koc indicates that it could go either way.
- BCF and Kow indicate that bioaccumulation should occur.

Exposure Analysis

WS indicates that this chemical should not leach and should adsorb to soil's organic matter, and it could runoff with soil. This is supported by values for BCF and Kow, which indicate bioaccumulation. Koc indicates it could go either way, but WS, BCF, and Kow show the opposite. Thus, it is predicted that the chemical could persist in the environment and could bioaccumulate. Based on its chemical structure, the amino moiety, it is predicted to be a carcinogen. In fact, it is a questionable carcinogen.

#2. Allyl chloride or 3-chloro-1-propene: mol. wt. 76.53 g/mol, log Kow 0.96, WS 3.6 X 10^{03} mg/L (UT) & VP 3.4 X 10^{02} mm Hg at 20° C. This chemical is a suspected carcinogen.

Predictive Equation Analysis

$$\log BCF = 0.76 \log Kow - 0.23 \qquad \text{(eq. 18.19)}$$
$$= 0.76 \, (\log Kow \; 0.96) - 0.23$$
$$= 0.4996$$

BCF $\quad = 3.1594$

$$\log Koc = -0.557 \log S + 4.277 \qquad \text{(eq. 18.16)}$$
$$= -0.557 \, (\log S \; 4.6727) + 4.277$$
$$= 1.674$$

Koc $\quad = 47.24$

Kow $\quad = 9.12$

Fate and Transport Analysis

- WS indicates that this chemical could leach, runoff, and be biodegraded.
- BCF and Kow indicate that bioaccumulation should not occur.
- It may volatilize and phototransformation may occur.

Exposure Analysis

This chemical could leach, runoff, and be biodegraded, and it should not be persistent in the environment. This is supported by the BCF and Kow, which indicate that bioaccumulation should not occur. This chemical may be volatile and phototransformation may occur; thus, inhalation of the chemical may be a problem, as it is a suspected carcinogen.

#3.   Aniline or benzenamine: mol. wt. 93.13 g/mol, log Kow 0.90, WS 3.5 X 10^4 mg/L & VP 1 X 10^{00} mm Hg at 35° C. This chemical is a suspected carcinogen.

Predictive Equation Analysis

log BCF = 0.76 log Kow − 0.23   (eq. 18.19)
       = 0.76(log Kow 0.90) − 0.23
       = 0.454
BCF   = 2.8

log Koc = 0.937 log Kow − 0.006   (eq. 18.17)
       = 0.937 (0.90) − 0.006
       = 0.837
Koc   = 6.875

Kow   = 7.943

Fate and Transport Analysis

- WS indicates that this chemical could leach, runoff, and be biodegraded.
- Koc indicates that it should not accumulate in soil.
- BCF and Kow indicate that it should not bioaccumulate.
- It may be volatile and phototransformation may occur.

Exposure Analysis

WS indicates that this chemical could leach, runoff, and be biodegraded. This is supported by Koc, which indicates that soil adsorption and soil accumulation should not occur. BCF and Kow indicate that bioaccumulation should not occur and this is supported by the WS value. It may volatilize and phototransformation may occur. Inhalation may be a problem, as it is, in fact, a suspected carcinogen. It can be predicted to be a carcinogen, based on the aniline or amine part of its chemical structure.

#4.   p-Chloroaniline: mol. wt. 127.57 g/mol, log Kow 1.83, WS 3.9×10^{03} mg/L 20 to 25° C & VP 1.5×10^{-02} mm Hg at 20° C. This chemical is questionable carcinogen.

Predictive Equation Analysis

$$\log BCF = 0.76 \log Kow - 0.23 \qquad \text{(eq. 18.19)}$$
$$= 0.76(\log Kow\ 1.83) - 0.23$$
$$= 1.1608$$
$$BCF = 14.48$$

$$\log Koc = 0.937 \log Kow - 0.006 \qquad \text{(eq. 18.17)}$$
$$= 0.937\ (1.83) - 0.006$$
$$= 1.709$$
$$Koc = 51.13$$

$$Kow = 67.608$$

Fate and Transport Analysis

• WS indicates that this chemical could leach, runoff, and be biodegraded.
• Koc indicates it should not accumulate in soil.
• BCF and Kow indicate that it should not bioaccumulate.
• It may be volatile and photodegradation may occur.

Exposure Analysis

WS indicates that this chemical could leach, runoff, and be biodegraded. This is supported by its Koc, which indicates that it should not accumulate in soil. Kow and BCF indicate that it should not bioaccumulate and this is supported by its WS. This chemical may be volatile and inhalation may be a problem. It is predicted to be a carcinogen, based on the aniline structure. In fact, it is a questionable carcinogen.

#5. Bis(2-Chloro-1-methylethyl)ether: mol. wt. 171.07 g/mol, log Kow 2.58, WS 1.7 X 10^{03} mg/L (UT) & VP 8.5 X 10^{-01} mm Hg at 20° C. This chemical is a questionable carcinogen.

Predictive Equation Analysis

log BCF= 0.76 log Kow – 0.23		(eq. 18.19)
= 0.76(log Kow 2.58) – 0.23		
= 1.7308		
BCF = 53.8		
log Koc = –0.557 log S + 4.277		(eq. 18.16)
= –0.557 (log S 3.9973) + 4.277		
= 2.0505		
Koc = 112.336		
Kow = 380.189		

Fate and Transport Analysis

• WS indicates that this chemical could leach, runoff, and be biode-graded.
• Koc indicates that soil accumulation should not occur.
• BCF and Kow indicate that it should not bioaccumulate.
• It may be volatile and phototransformation may occur.

Exposure Analysis

WS indicates that this chemical could leach, runoff, and be biodegraded. This is supported by its Koc value, which indicates that adsorption to soil's organic matter should occur. BCF and Kow indicate that bioaccumulation should not occur and this is supported by WS value. It may be volatile and inhalation may be a problem, as it is questionable carcinogen.

#6. Chloroprene or 2-chloro-1,3-butadiene or Neoprene: mol. wt. 88.54 g/mol, log Kow 0.57, & VP 2×10^{02} mm Hg at 20° C. This chemical is a questionable carcinogen.

Predictive Equation Analysis

log BCF =	$0.76 \log Kow - 0.23$	(eq. 18.19)
=	$0.76(\log Kow\ 0.57) - 0.23$	
=	0.203	
BCF	$= 1.597$	

log Koc =	$-0.557 \log S + 4.277$	(eq. 18.16)
=	$-0.557\ (\log S\ 6.4791) + 4.277$	
=	0.6611	
Koc	$= 4.5829$	

log S	$= -1.37 \log Kow + 7.26$	(eq. 18.1)
	$= -1.37(0.57) + 7.26$	
	$= 6.4791$	
S	$= 3,013,699.87\ \mu\ mol/L$	
S	$= 266,832.9869\ mg/L$	
Kow	$= 3.7154$	

Fate and Transport Analysis

• WS indicates that this chemical could leach, runoff, and be biode-graded.
• Koc indicates that it should not accumulate in soil.
• Kow and BCF indicate that it should not bioaccumulate.
• It may be volatile and phototransformation may occur.

Exposure Analysis

WS indicates that it may leach, runoff, and be biodegraded. This is supported by the Koc value, which indicates that it should not be adsorbed to soil's organic matter. BCF and Kow indicate that it should not bioaccumulate, and this is supported by the WS value. It may be volatile and inhalation may be a problem, as this chemical is a question-able carcinogen.

#7. m-Cresol or 3-methylphenol: mol. wt. 108.14 g/mol, log Kow 1.96, WS 2.5 X 10^{04} mg/L (UT), & VP 1.1 X 10^{-01} mm Hg at 20° C. This chemical is a questionable carcinogen.

Predictive Equation Analysis

$$\log BCF = 0.76 \log Kow - 0.23 \qquad \text{(eq. 18.19)}$$
$$= 0.76(\log Kow\ 1.96) - 0.23$$
$$= 1.2596$$

BCF = 18.18

Kow = 91.2

Fate and Transport Analysis

- WS indicates that this chemical could leach, runoff, and be biode-graded.
- BCF and Kow indicate that is should not bioaccumulate.
- VP indicates that it may be volatile and phototransformation may occur.

Exposure Analysis

WS indicates that this chemical could leach, runoff, and be biodegraded, and this is supported by Kow and BCF values. Kow and BCF indicate that it should not bioaccumulate. It may be volatile and inhalation may be a problem, as it is a questionable carcinogen.

#8. Diallate or S-(2,3-dichloroally) diisopropylthiocarbamate: mol. wt. 270.24 g/mol, log Kow 4.49, WS 1.4 X 10^{01} mg/L, & VP 1.5 X 10^{-04} mm Hg at 25° C. This chemical is a questionable carcinogen.

Predictive Equation Analysis

$$\log BCF = 0.76 \log Kow - 0.23 \qquad \text{(eq. 18.19)}$$
$$= 0.76(\log Kow\ 4.49) - 0.23$$
$$= 3.182$$

BCF = 1,521.948

$$\log Koc = -0.557 \log S + 4.277 \qquad \text{(eq. 18.16)}$$
$$= -0.557\ (\log S\ 1.7144) + 4.277$$
$$= 3.3221$$

Koc = 2,099.32

Kow = 30,902.95

Fate and Transport Analysis

- WS indicates that this chemical could leach, runoff, and be biodegraded, the value is close and it may go the other way.
- Koc indicates that it could accumulate in soil.
- BCF and Kow indicate that it could bioaccumulate.

Exposure Analysis

WS indicates that this chemical could leach, runoff, and be biodegraded. Data are contradictive in that Koc value indicates that is should be adsorbed to soil's organic matter, and BCF and Kow indicate that it could bioaccumulate. The WS is close to <10; thus, it could go either way. It is better to take the safe side and say it should not leach and should runoff with soil particles and accumulate in soil. It may be volatile and inhalation may be a problem, as it is a questionable carcinogen.

#9. Dibenzofuran: mol. wt. 168.19 g/mol, log Kow 4.12, & WS 1 X 10^{01} mg/L (UT). This chemical is a questionable carcinogen.

Predictive Equation Analysis

$$\log BCF = 0.76 \log Kow - 0.23 \qquad \text{(eq. 18.19)}$$
$$= 0.76(\log Kow\ 4.49) - 0.23$$
$$= 3.182$$

BCF $= 1,521.948$

Kow $= 13,182.567$

Fate and Transport Analysis

- WS indicates that this chemical should not leach, could runoff with soil particles, and should not be biodegraded.
- BCF and Kow indicate that it could bioaccumulate.

Exposure Analysis

WS indicates that this chemical should not leach and should not be biodegraded. It is predicted to accumulate in soil's organic matter. Thus, it could runoff with soil matter. BCF and Kow values support this prediction. BCF and Kow indicate that it could bioaccumulate and this is supported by it WS value. It is a questionable carcinogen and it may be a problem if inhaled; however there is no volatility data to make this prediction.

#10. Di-n-butyl phthalate: mol. wt. 278.35 g/mol, log Kow 5.2 & WS 4 X 10^{02} mg/L at 25° C.

Predictive Equation Analysis

$$\log BCF = 0.76 \log Kow - 0.23 \qquad \text{(eq. 18.19)}$$
$$= 0.76(\log Kow\ 5.2) - 0.23$$
$$= 3.722$$
$$BCF = 5{,}272.299$$
$$Kow = 158{,}489.319$$

Fate and Transport Analysis

• WS indicates that this chemical should not leach, may runoff soil with soil particles, and may accumulate in soil.
• BCF and Kow indicate that it should bioaccumulate.

Exposure Analysis

WS indicates that this chemical should not leach and could accumulate in soil. This is supported by the Kow value. BCF and Kow indicate that it could bioaccumulate and this is supported by the WS value. Many phthalates are questionable carcinogens; thus, based on this, it is predicted to be a carcinogen.

#11. 2,3-Dichlorobenzidine: mol. wt. 253.13 g/mol, log Kow 3.51, WS 3.1 X 10^{00} mg/L at 25° C. This chemical is a confirmed carcinogen.

Predictive Equation Analysis

$$\log BCF = 0.76 \log Kow - 0.23 \qquad \text{(eq. 18.19)}$$
$$= 0.76(\log Kow\ 3.51) - 0.23$$
$$= 2.4376$$
$$BCF = 273.905$$
$$Kow = 3{,}235.936$$

Fate and Transport Analysis

- WS indicates that this chemical should not leach, may runoff with soil particles, and should not be biodegraded.
- BCF and Kow indicate that is should be bioaccumulated.

Exposure Analysis

WS indicates that this chemical should not leach and could be accumulated in soil's organic matter. This is supported by the Kow value. Kow and BCF indicate that it should bioaccumulate and this is supported by the WS value. This chemical is predicted to be a carcinogen. If volatile, inhalation could be a problem, as this chemical is, in fact, a confirmed carcinogen.

#12. 2,6-Dichlorophenol: mol. wt. 163.00 g/mol, & log Kow 2.88.

Predictive Equation Analysis

$$\log BCF = 0.76 \log Kow - 0.23 \qquad \text{(eq. 18.19)}$$
$$= 0.76(\log Kow\ 2.88) - 0.23$$
$$= 1.9588$$

$BCF \quad = 90.94$

$$\log S \quad = -1.37 \log Kow + 7.26 \qquad \text{(eq. 18.1)}$$
$$= -1.37\ (2.88) + 7.26$$
$$= 3.3144$$

$S \quad = 2{,}062.5287\ \mu\ mol/L$

$S \quad = 336.19\ mg/L$

$Kow \quad = 758.5776$

Fate and Transport Analysis

- WS indicates that this chemical should leach, runoff, and be biodegraded.
- BCF and Kow indicate that bioaccumulation could go either way.

Exposure Analysis

WS indicates that this chemical could leach, runoff, and be biodegraded; however, other predictive data indicates that it could go either way. BCF and Kow are on the high side. Although data indicates that bioaccumulation could go either way, it may be best to error on the safe side and predict that it could bioaccumulate. The predictive data do not give a clear picture and real data are needed.

#13. 1,2-Dichloropropane: mol. wt. 112.99 g/mol, log Kow 2.28, WS 2.7 X 10^{03} mg/L at 20° C. This chemical may breakdown to phosgene, which is a toxic gas.

Predictive Equation Analysis

$$\begin{aligned} \log BCF &= 0.76 \log Kow - 0.23 \quad\quad\quad (eq.\ 18.19) \\ &= 0.76(\log Kow\ 2.28) - 0.23 \\ &= 1.5028 \end{aligned}$$

BCF   = 31.827

Kow   = 190.546

Fate and Transport Analysis

• WS indicates that this chemical could leach, runoff, and be biodegraded.
• BCF and Kow indicate that bioaccumulation should not occur.

Exposure Analysis

WS indicates that this chemical could leach and runoff. It also indicates that it could biodegrade, and this is supported by the BCF and Kow values. This chemical may breakdown (be degraded) to phosgene, a poisonous gas. Inhalation could be a problem if such degradation occurs and if exposure to phosgene gas occurs. Since the chemical following this chemical is a confirmed carcinogen, and these two chemicals are similar, it is predicted that this chemical may also be a carcinogen.

#14. cis-1,3-Dichloropene: mol. wt. 110.97 g/mol, log Kow 1.41, WS 2.7 X 10^{03} mg/L (UT), & VP 4.3 X 10^{01} mm Hg (UT). This chemical is a confirmed carcinogen.

Predictive Equation Analysis

$$\begin{aligned} \log BCF &= 0.76 \log Kow - 0.23 \quad\quad\quad (eq.\ 18.19) \\ &= 0.76(\log Kow\ 1.41) - 0.23 \\ &= 0.8416 \end{aligned}$$

BCF   = 6.9438

$$\log K_{oc} = -0.557 \log S + 4.277 \qquad \text{(eq. 18.16)}$$
$$= -0.557 (2.476) + 4.277$$
$$= 2.8976$$

$K_{oc} = 789.86$

$K_{ow} = 25.7046$

Fate and Transport Analysis

- WS indicates that this chemical could leach, runoff, and be biodegraded.
- Koc indicates that accumulation should not occur.
- BCF and Kow indicate that bioaccumulation should not occur.
- It may be volatile and phototransformation may occur.

Exposure Analysis

WS indicates that this chemical could leach and runoff. It also indicates that biodegradation could occur and this is supported by the Koc value. Koc indicates that accumulation should not occur. BCF and Kow indicate that biodegradation should not occur. If it volatilizes, than inhalation could be a problem, as this chemical is a confirmed carcinogen. Since it is structurally similar to the chemical given above, care should be taken in that it may be degraded to phosgene, a toxic gas.

#15. p-(Dimethylamino) azobenzene: mol. wt. 225.28 g/mol, log Kow 4.58, & WS insoluble. This chemical is a confirmed carcinogen.

Predictive Equation Analysis

$$\log BCF = 0.76 \log K_{ow} - 0.23 \qquad \text{(eq. 18.19)}$$
$$= 0.76(\log K_{ow}\ 4.58) - 0.23$$
$$= 3.2508$$

$BCF = 1,781.558$

$$\log S = -1.37 \log K_{ow} + 7.26 \qquad \text{(eq. 18.1)}$$
$$= -1.37 (4.58) + 7.26$$
$$= 0.9854$$

$S = 9.6694\ \mu\ mol/L$

$S = 2.178\ mg/L$

$K_{ow} = 38,018.9396$

Fate and Transport Analysis

- WS indicates that this chemical should not leach, could runoff with soil particles, and should not be biodegraded.
- BCF and Kow indicate that it should bioaccumulate.

Exposure Analysis

WS indicates that this chemical should not leach and should not be biodegraded. This is supported by the Kow and BCF values. It may runoff with soil matter. BCF and Kow indicate that it could bioaccumulate and this is supported by WS value. Based on chemical structure, this chemical has an azo and amino type moiety, thus, it is predicted to be a carcinogen. In fact, it is a confirmed carcinogen.

#16. 2,4-Dimethylphenol or m-xylenol: mol. wt. 122.19 g/mol, log Kow 2.42 & WS 7.868×10^{02} mg/L at 25° C. This chemical is a questionable carcinogen.

Predictive Equation Analysis

$$\log BCF = 0.76 \log Kow - 0.23 \qquad \text{(eq. 18.19)}$$
$$= 0.76 (2.42) - 0.23$$
$$= 1.609$$

BCF $= 40.66$

Kow $= 263$

Fate and Transport Analysis

- WS indicates that this chemical could leach, runoff, and be biodegraded.
- BCF and Kow indicate that it should not bioaccumulate.

Exposure Analysis

WS indicates that this chemical could leach, runoff, and be biodegraded. This is supported in that Kow and BCF values indicate that it should not accumulate. This chemical is a questionable carcinogen and exposure by ingestion should be prevented. It is not known if it is volatile; if shown to be volatile, however, inhalation should be prevented, as it is, in fact, a questionable carcinogen.

#17. 1,3-Dinitrobenzene: mol. wt. 168.11 g/mol, log Kow 1.49, WS 4.69 X 10^{02} at 15° C & VP 8.15 X 10^{-04} mm Hg at 35° C. This chemical is a suspected carcinogen.

Predictive Equation Analysis

$$\log BCF = 0.76 \log Kow - 0.23 \qquad \text{(eq. 18.19)}$$
$$= 0.76\,(1.49) - 0.23$$
$$= 0.9024$$

BCF   = 7.987

Kow   = 30.903

Fate and Transport Analysis

- WS indicates that this chemical could leach, runoff, and be biodegraded.
- BCF and Kow indicate that it should not bioaccumulate.
- VP indicates that it may be volatile and phototransformation may occur.

Exposure Analysis

WS indicates that this chemical could leach, runoff, and be biodegraded. This is supported by the BCF and Kow values, which indicate that it should not bioaccumulate. VP indicates that it may be volatile; thus, inhalation may be a problem, as it is a suspected carcinogen.

#18. Di-n-octylphthalate: mol. wt. 390.57 g/mol, log Kow 9.2 & WS 3 X 10^{00} at 25° C.

Predictive Equation Analysis

$$\log BCF = 0.76 \log Kow - 0.23 \qquad \text{(eq. 18.19)}$$
$$= 0.76\,(9.2) - 0.23$$
$$= 6.762$$

BCF   = 5,780,960.474

Kow   = 1,584,893,192

Fate and Transport Analysis

- WS indicates that this chemical should not leach, could runoff with soil particles, and could accumulate in soil.

- Kow and BCF indicate that this chemical should bioaccumulate.

Exposure Analysis

Kow and BCF indicate that this chemical should bioaccumulate. The values for Kow and BCF also indicate that the chemical should not leach, could runoff with soil matter, and could accumulate in soil's organic matter. This is supported by the WS value. Many phthalates are confirmed carcinogens; thus, it is predicted to be a carcinogen.

#19. Endrin aldehyde: mol. wt. 380.92 g/mol, log Kow 5.6, WS 2.6 X 10^{-01} mg/L at 25° C & VP 2 X 10^{-07} mm Hg at 25° C. Endrin is a questionable carcinogen.

Predictive Equation Analysis

$$\log BCF = 0.76 \log Kow - 0.23 \qquad \text{(eq. 18.19)}$$
$$= 0.76\,(5.6) - 0.23$$
$$= 4.026$$

BCF $= 10{,}616.956$

Kow $= 398{,}107.1706$

Fate and Transport Analysis

- WS indicates that this chemical should not leach, could runoff with soil particles, and could accumulate in soil's organic matter.
- BCF and Kow indicate that it should bioaccumulate.
- VP indicates that it may or may not be volatile.

Exposure Analysis

WS indicates that this chemical should not leach, could runoff, and should accumulate in soil's organic matter. This supported by Kow and BCF values, which indicate that it should bioaccumulate. VP indicates that volatility may or may not be a problem; however, volatility cannot be predicted with the data. This chemical is an aldehyde type chemical, and it is predicted to be a carcinogen. Its parent compound, Endrin, is a questionable carcinogen; thus, it may be better to error on the side of safety and predict it may be volatile.

#20. Bis(2-Ethylhexyl) phthalate: mol. wt. 390.54 g/mol, log Kow 5.3, & WS 4 X 10^{-01} mg/L at 25° C. This chemical is a confirmed carcinogen.

Predictive Equation Analysis

log BCF = 0.76 log Kow − 0.23   (eq. 18.19)
        = 0.76 (5.3) − 0.23
        = 379.8
BCF   = 16,280.58
Kow   = 199,526.232

Fate and Transport Analysis

- WS indicates that this chemical should not leach, could runoff with soil matter, and should accumulate in soil.
- BCF and Kow indicate that it could bioaccumulate.

Exposure Analysis

WS indicates that this chemical should not leach, could runoff with soil matter, and should accumulate in soil's organic matter. This is supported by the Kow and BCF values indicating that it should bioaccumulate. Based on the chemical structure, and knowing that many phthalates are carcinogens, it is predicted to be a carcinogen. In fact, it is a confirmed carcinogen.

#21. Hexachlorophene: mol. wt. 406.91 g/mol, log Kow 7.54, & WS 4 X 10^{-03} mg/L (UT). This chemical is a carcinogen.

Predictive Equation Analysis

log BCF = 0.76 log Kow − 0.23   (eq. 18.19)
        = 0.76 (7.54) − 0.23
        = 5.5004
BCF   = 316,519.1567
Kow   = 34,673,685.04

Fate and Transport Analysis

- WS indicates that this chemical should not leach, could runoff with soil particles, and could accumulate in soil.

- BCF and Kow indicate that it is lipo soluble and could bioaccumulate.

Exposure Analysis

WS indicates that this chemical should not leach, could runoff with soil matter, and should accumulate in soil's organic matter. This is supported by the high values for Kow and BCF. Kow and BCF indicate that this chemical should bioaccumulate and food chain contaminations could be expected; this may be a problem, as it is a confirmed carcinogen

#22. Indeno(1,2,3-cd)pyrene: mol. wt. 276.34 g/mol, log Kow 7.66, WS 6.2 X 10^{-02} mg/L (UT), & VP 1 X 10^{-10} mm Hg at 25° C. This chemical is a confirmed carcinogen.

Predictive Equation Analysis

$$\log BCF = 0.76 \log Kow - 0.23 \quad\quad\quad (eq. 18.19)$$
$$= 0.76\,(7.66) - 0.23$$
$$= 5.5916$$

BCF $\quad = 390,481.0837$

Kow $\quad = 45,708,818.96$

Fate and Transport Analysis

- WS indicates that this chemical should not leach, could runoff with soil particles, and should accumulate in soil.
- BCF and Kow indicate that it could bioaccumulate.
- VP indicates that it should not be volatile.

Exposure Analysis

WS indicates that this chemical should not leach, could runoff with soil matter, and should accumulate to soil's organic matter. This is supported by high the Kow and BCF values. BCF and Kow indicate that it could bioaccumulate, and food chain contamination could be a problem, as this chemical is a confirmed carcinogen. Volatility should not occur; thus, inhalation should not be a problem unless it sublimes.

#23. Isophorone: mol. wt. 138.21 g/mol, log Kow 1.7, WS 1.2 X 10^{04} mg/L (UT), & VP 3.8 X 10^{-01} mm Hg at 20° C. This chemical is a questionable carcinogen.

Predictive Equation Analysis

$$\log BCF = 0.76 \log Kow - 0.23 \qquad \text{(eq. 18.19)}$$
$$= 0.76 (1.7) - 0.23$$
$$= 1.062$$

BCF $= 11.5345$

Kow $= 50.1187$

Fate and Transport Analysis

- WS indicates that this chemical could leach, runoff, and be biodegraded.
- Kow and BCF indicate that it should not bioaccumulate.
- VP indicates it may be volatile and phototransformation may occur.

Exposure Analysis

WS indicates that this chemical could leach, runoff, and be degraded by microorganisms. This is supported by the Kow and BCF values. BCF and Kow indicate that it should not bioaccumulate or cause food chain contamination. VP indicates that volatility is likely and inhalation may be a problem, as it is a questionable carcinogen.

#24. Isopropylbenzene or cumene: mol. wt. 120.19 g/mol, log Kow 3.66, WS 5 X 10^{01} mg/L at 20° C, & VP 3.2 X 10^{00} mm Hg at 20° C.

Predictive Equation Analysis

$$\log BCF = 0.76 \log Kow - 0.23 \qquad \text{(eq. 18.19)}$$
$$= 0.76 (3.66) - 0.23$$
$$= 2.5516$$

BCF $= 356.123$

Kow $= 4,570.8819$

Fate and Transport Analysis

- WS indicates that this chemical should not leach, could runoff with soil particles, and could accumulate in soil.

- BCF and Kow indicate that it could bioaccumulate.
- VP indicates that it could be volatile and photodegradation may occur.

Exposure Analysis

WS indicates that this chemical should not leach and could runoff with soil matter. It also indicates that it should accumulate in soil's organic matter and this is supported by the values for BCF and Kow. BCF and Kow indicate that this chemical could bioaccumulate and food chain contamination is likely. VP indicates that volatility could occur and inhalation may be a problem. Based on its chemical structure, it is predicted that this chemical is a carcinogen, especially if the isopropyl moiety is broken off and benzene is formed.

#25. p-Isopropyltoluene: mol. wt. 134.22 g/mol, log Kow 4.1, WS 3.4 X 10^{02} mg/L (UT), & VP 1 X 10^{00} mm Hg at 17.3° C.

Predictive Equation Analysis

$$\log BCF = 0.76 \log Kow - 0.23 \qquad \text{(eq. 18.19)}$$
$$= 0.76 (4.1) - 0.23$$
$$= 2.886$$
$$BCF = 769.13$$
$$Kow = 12,589.254$$

Fate and Transport Analysis

- WS indicates that this chemical could leach, runoff, and be biodegraded, but it could go either way.
- BCF and Kow indicate that it should bioaccumulate.
- VP indicates that it may be volatile and phototransformation may occur.

Exposure Analysis

WS indicates that this chemical could leach, runoff, and be biodegraded, but it could go either way. However, its value is nearer to the side that it should not leach, could runoff with soil matter, and could accumulate in soil's organic matter. This is supported by the values for BCF and Kow. BCF and Kow indicate that it could bioaccumulate and food chain contamination is possible. VP indicates it may volatilize; thus, inhalation may be a problem.

#26. Isosafrole: mol. wt. 162.19 g/mol, log Kow 2.66, WS 1.09 X 10^{03} mg/L (UT), & VP 1.6 X 10^{-08} mm Hg (UT). This chemical is a questionable carcinogen.

Predictive Equation Analysis

$$\begin{aligned} \log BCF &= 0.76 \log Kow - 0.23 \qquad \text{(eq. 18.19)} \\ &= 0.76\,(2.66) - 0.23 \\ &= 1.7916 \\ BCF &= 61.887 \\ Kow &= 457.088 \end{aligned}$$

Fate and Transport Analysis

- WS indicates that this chemical could leach, runoff, and be biodegraded.
- BCF and Kow indicate that it should not bioaccumulate.
- VP indicates that it should not volatilize.

Exposure Analysis

WS indicates that this chemical could leach, runoff, and be biodegraded. This is supported by the values for BCF and Kow. Kow and BCF indicate that bioaccumulation should not occur; thus, food chain contamination is not likely. VP indicates that it should not volatilize; thus, inhalation should not be a problem. This chemical contains the safrole moiety; therefore, it is predicted to be a carcinogen. In fact, it is a questionable carcinogen; therefore, ingestion of this chemical may be a problem.

#27. Methapyrilene or N,N-dimethyl-N'-2-pyridinyl-N'-(2-thienylme-thyl)-1,2-ethanediamine: mol. wt. 261.38 g/mol, & log Kow 2.74.

Predictive Equation Analysis

$$\begin{aligned} \log BCF &= 0.76 \log Kow - 0.23 \qquad \text{(eq. 18.19)} \\ &= 0.76\,(2.74) - 0.23 \\ &= 1.8524 \\ BCF &= 71.1869 \end{aligned}$$

$$\log 1/S = 0.996 \log Kow - 0.339 \text{ (mol/L)} \qquad \text{(eq. 18.10)}$$
$$= 0.996 (2.74) - 0.339$$
$$= 2.39$$

S   = 0.4184 mol/L

S   = 109,362.186 mg/L

Kow = 549.5409

Fate and Transport Analysis

• WS indicates that this chemical could leach, runoff, and be biode-graded.
• BCF and Kow indicate that it should not bioaccumulate; however, Kow value indicates it could go either way.

Exposure Analysis

WS indicates that this chemical could leach, runoff, and be biodegraded. This is supported by the values for BCF and Kow. BCF and Kow indicate that bioaccumulation should not occur and food chain contamination is unlikely. This chemical is an amine/diamine and it is predicted to be a carcinogen.

#28. 3-Methylcholanthrene: mol. wt. 268.34 g/mol, log Kow 7.11, & WS 3 X 10⁻⁰³ mg/L at 25° C. This chemical is a suspected carcinogen.

Predictive Equation Analysis

$$\log BCF = 0.76 \log Kow - 0.23 \qquad \text{(eq. 18.19)}$$
$$= 0.76 (7.11) - 0.23$$
$$= 5.1736$$

BCF   = 149,142.0128

Kow   = 12,882,495.51

Fate and Transport Analysis

• WS indicates that this chemical should not leach, could runoff with soil particles, and should accumulate in soil.
• BCF and Kow indicate that it should bioaccumulate.

Exposure Analysis

WS indicates that this chemical should not leach, could runoff, and should accumulate in soil's organic matter. This is supported by the high values for BCF and Kow. BCF and Kow indicate that it could bioaccumulate and food chain contamination is likely.

#29. Methylethylketone or MEK: mol. wt. 72.11 g/mol, log Kow 0.26, WS 2.75 X 10^{05} mg/L (UT), & VP 1 X 10^{02} mm Hg at 25° C.

Predictive Equation Analysis

$$\log BCF = 0.76 \log Kow - 0.23 \qquad \text{(eq. 18.19)}$$
$$= 0.76\,(0.26) - 0.23$$
$$= -0.0324$$

BCF $= 0.928$

$$\log 1/S = 1.229 \log Kow - 0.72 \text{ (mol/L)} \qquad \text{(eq. 18.3)}$$
$$= 1.229\,(0.26) - 0.72$$
$$= -0.4005$$

1/S $= 0.3976$
S $= 2.5148$ mol/L
S $= 181,340.7867$ mg/L

Kow $= 1.8197$

Fate and Transport Analysis

- WS indicates that this chemical could leach, runoff, and be biode-graded.
- BCF and Kow indicate that is should not bioaccumulate.
- VP indicates that it may volatilize and photodegradation may occur.

Exposure Analysis

WS indicates that this chemical could leach, runoff, and be biodegraded. This is supported by the values for Kow and BCF, which indicates that it should not bioaccumulate. It may volatilize; thus, inhalation may be a problem.

#30. Methyl iodine or iodomethane: mol. wt. 141.94 g/mol, log Kow 1.69, WS 1.4 X 10^{04} mg/L at 20° C, & VP 4 X 10^{02} mm Hg at 25.3° C. This chemical is a confirmed carcinogen.

Predictive Equation Analysis

log BCF = 0.76 log Kow – 0.23   (eq. 18.19)
 = 0.76 (1.69) – 0.23
 = 1.0544
BCF = 11.3344
Kow = 48.9779

Fate and Transport Analysis

- WS indicates that this chemical could leach, runoff, and be biodegraded.
- BCF and Kow indicate that it should not bioaccumulate.
- VP indicates that it may be volatile and photodegradation may occur.

Exposure Analysis

WS indicates that this chemical could leach, runoff, and be biodegraded. This is supported by the BCF and Kow values, which indicate that bioaccumulation should not occur. VP indicates volatilization may occur; thus, inhalation may be a problem, as this chemical is, in fact, a confirmed carcinogen.

#31. Methyl methacrylate: mol. wt. 100.12 g/mol, log Kow 1.33, WS 1.6 X 10^{04} mg/L (UT), & VP 2.8 X 10^{01} mm Hg at 20° C. This chemical is a questionable carcinogen.

Predictive Equation Analysis

log BCF = 0.76 log Kow – 0.23   (eq. 18.19)
 = 0.76 (1.33) – 0.23
 = 0.7808
BCF = 6.0367
Kow = 21.3796

Fate and Transport Analysis

- WS indicates that this chemical could leach, runoff, and be biodegraded.
- BCF and Kow indicate that it should not bioaccumulate.
- VP indicates that it may volatilize and phototransformation may occur.

Exposure Analysis

WS indicates that this chemical could leach, runoff, and be biodegraded. This is supported by the values for BCF and Kow, which indicate that it should not bioaccumulate. VP indicates that it may volatilize; thus, inhalation may be a problem, as this it is a questionable carcinogen.

#32. Methyl methanesulfonate: mol. wt. 110.13 g/mol, & WS 2×10^{05} mg/L (UT). This chemical is a questionable carcinogen.

Predictive Equation Analysis

$\log 1/S = 1.013 \log Kow - 0.52$ (mol/L)   (eq. 18.4)
$\log 1/200{,}000$ mg/L $= 1.013 \log Kow - 0.52$
$\log 1/1.8132$ mol/L $+ 0.52 = 1.013 \log Kow$
$(\log 1.0715$ mol/L$)/1.013 = \log Kow$
$\log 1.0577 = \log Kow$
$\log Kow = 1.0577$

Kow $= 11.4222$

$\log BCF = 0.76 \log Kow - 0.23$ (eq. 18.19)
$= 0.76 (1.0577) - 0.23$
$= 0.5739$
BCF $= 3.7485$

Fate and Transport Analysis

- WS indicates that this chemical could leach, runoff, and be biodegraded.
- BCF and Kow indicate that it should not bioaccumulate.

Exposure Analysis

WS indicates that this chemical could leach, runoff, and be biodegraded. This is supported by the Kow and BCF predicted values, which indicate that it should not bioaccumulate. This chemical is a questionable carcinogen; thus, exposure to it may be a problem.

#33. Methyl isobutylketone: mol. wt. 100.16 g/mol, log Kow 1.09, WS 1.91 X 10^{04} mg/L (UT), & VP 1.0 X 10^{01} mm Hg at 30° C.

Predictive Equation Analysis

log BCF = 0.76 log Kow – 0.23   (eq. 18.19)
$\qquad$ = 0.76 (1.09) – 0.23
$\qquad$ = 0.0.5984
BCF $\qquad$ = 3.9664
Kow $\qquad$ = 12.3027

Fate and Transport Analysis

• WS indicates that this chemical could leach, runoff, and be biodegraded.
• BCF and Kow indicate that it should not bioaccumulate.
• VP indicates that it may be volatile and photodegradation may occur.

Exposure Analysis

WS indicates that this chemical could leach, runoff, and be biodegraded. This is supported by the Kow and BCF values, which indicate that is should not bioaccumulate. It may be volatile; thus, inhalation may be a problem.

#34. Methyl tert-butyl ether or 2-methoxy-2-methylpropane: mol. wt. 88.15 g/mol, WS 4.8 X 10^{04} mg/L (UT), and VP 2.45 X 10^{02} mm Hg at 25° C.

Predictive Equation Analysis

log 1/S $\quad$ = 1.182 log Kow – 0.935 (mol/L)   (eq. 18.5)
log Kow = $\underline{\text{log } 1/-0.2640 + 0.935}$
$\qquad\qquad$ 1.182
Kow $\qquad$ = 0.0039
WS $\qquad$ = 48,000 mg/L

Fate and Transport Analysis

- WS indicates that this chemical could leach, runoff, and be biodegraded.
- Kow indicates that it should not bioaccumulate.
- VP indicates that it may be volatile and phototransformation may occur.

Exposure Analysis

WS indicates that this chemical could leach, runoff, and be biodegraded. This is supported by the predicted Kow value, which indicates that it should not bioaccumulate. VP indicates that it may volatilize; thus, inhalation may be a problem.

#35. 1,4-Naphthoquinone: mol. wt. 158.16 g/mol, & log Kow 1.71/1.78. This chemical is a questionable carcinogen. Since there are two Kow values for this chemical, they both will be used.

Predictive Equation Analysis

$$\log BCF = 0.76 \log Kow - 0.23 \qquad \text{(eq. 18.19)}$$
$$= 0.76 (1.71) - 0.23$$
$$= 1.0696$$
$$BCF = 11.7382$$
$$Kow = 51.2861$$

$$\log BCF = 0.76 \log Kow - 0.23 \qquad \text{(eq. 18.19)}$$
$$= 0.76 (1.78) - 0.23$$
$$= 1.1228$$
$$BCF = 13.2678$$
$$Kow = 60.256$$

Fate and Transport Analysis

- Kow and BCF indicate that this chemical should not bioaccumulate.

Exposure Analysis

Kow and BCF indicate that this chemical should not bioaccumulate. Based on these values, it is predicted that it should not leach, could runoff with soil particles, and could accumulate in soil's organic matter. It is a questionable carcinogen; thus, exposure should be prevented.

#36. 1-Naphthylamine: mol. wt. 143.19 g/mol, log Kow 2.22, & WS 1.7 X 10^{03} mg/L (UT). This chemical is a confirmed carcinogen.

Predictive Equation Analysis

$$\log BCF = 0.76 \log Kow - 0.23 \qquad \text{(eq. 18.19)}$$
$$= 0.76\,(2.22) - 0.23$$
$$= 1.4572$$

BCF = 28.655

Kow = 165.9587

Fate and Transport Analysis

- WS indicates that this chemical could leach, runoff, and be biodegraded.
- BCF and Kow indicate that it should not bioaccumulate.

Exposure Analysis

WS indicates that this chemical could leach, runoff, and be biodegraded. This is supported by the BCF and Kow values, which indicate that it should not bioaccumulate. Because it is predicted to be a carcinogen, based on the amine moiety, exposure should be prevented. In fact, it is a confirmed carcinogen.

#37. 2-Naphthylamine: mol. wt. 143.19 g/mol, log Kow 2.25, & WS 5.86 X 10^{02} mg/L at 20-30° C. This chemical is a confirmed carcinogen.

Predictive Equation Analysis

$$\log BCF = 0.76 \log Kow - 0.23 \qquad \text{(eq. 18.19)}$$
$$= 0.76\,(2.25) - 0.23$$
$$= 1.48$$

BCF = 30.1995

Kow = 177.8279

Fate and Transport Analysis

- WS indicates that this chemical could leach, runoff, and be biodegraded.
- BCF and Kow indicate that it should not bioaccumulate.

Exposure Analysis

WS indicates that this chemical could leach, runoff, and be biodegraded. This is supported by the BCF and Kow values, which indicate that it should not bioaccumulate. Exposure should be prevented; it is predicted to be a carcinogen, based on the amine moiety. In fact, it is a confirmed carcinogen.

#38. 2-Nitroaniline: mol. wt. 138.13 g/mol, log Kow 1.83, WS 1.26 X 10^{03} mg/L at 25° C, & VP <1 X 10^{-01} mm Hg at 30° C.

Predictive Equation Analysis

$$\log BCF = 0.76 \log Kow - 0.23 \qquad \text{(eq. 18.19)}$$
$$= 0.76\,(1.83) - 0.23$$
$$= 1.1608$$

BCF $= 14.481$

Kow $= 67.6083$

Fate and Transport Analysis

- WS indicates that this chemical could leach, runoff, and be biodegrade.
- BCF and Kow indicate that it should not bioaccumulate.
- VP indicates that it may be volatile and phototransformation may occur.

Exposure Analysis

WS indicates that this chemical could leach, runoff, and be biodegraded. This is supported by the BCF and Kow values, which indicate that it should not bioaccumulate. VP indicates that it may be volatile and inhalation may be a problem, as it is predicted to be a carcinogen, based on the aniline moiety.

#39. 4-Nitroquinoline-1-oxide: mol. wt . 190.16 g/mol. No other data. This chemical is a suspected carcinogen.

Predictive Equation Analysis

No data to put into equations.

Fate and Transport Analysis

No data to predict.

Exposure Analysis

The only prediction that can be made about this chemical is that it may be a carcinogen, based on the nitro and benzene moieties. In fact, it is a suspected carcinogen.

#40. N-Nitrosodibutylamine: mol. wt. 158.24 g/mol. This chemical is a confirmed carcinogen.

Predictive Equation Analysis

No data to put into equations.

Fate and Transport Analysis

No data to predict.

Exposure Analysis

Based on the chemical structure, this chemical contains an amine and nitroso moiety; thus, it is predicted to be a carcinogen. In fact, it is a confirmed carcinogen.

#41. N-Nitrosodiethylamine: mol. wt. 102.14 g/mol, & soluble. This chemical is a confirmed carcinogen.

Predictive Equation Analysis

No data to put into equations.

Fate and Transport Analysis

• It is stated to be soluble; thus, it could leach, runoff, be biodegraded, and should not bioaccumulate.

Exposure Analysis

This chemical is soluble and could be mobile in the environment, but should be biodegraded over time. It should not bioaccumulate. Based on the chemical structure, this chemical contains an amine and nitroso moiety and is predicted to be a carcinogen. In fact, it is a confirmed carcinogen.

#42. N-Nitrosodimethylamine: mol. wt. 74.08 g/mol, log Kow –0.47, WS miscible, & VP 8 X 10^{00} mm Hg at 25° C. This chemical is a confirmed carcinogen.

Predictive Equation Analysis

$$\log BCF = 0.76 \log Kow - 0.23 \qquad \text{(eq. 18.19)}$$
$$= 0.76\,(-0.47) - 0.23$$
$$= -0.5872$$

BCF = 0.2587

Kow = 0.3388

Fate and Transport Analysis

- WS indicates that this chemical could leach, runoff, and be biodegraded.
- BCF and Kow indicate that it should not bioaccumulate.
- VP indicates that it may be volatile and phototransformation may occur.

Exposure Analysis

WS indicates that this chemical could leach, runoff, and be biodegraded. This is supported by the BCF and Kow values, which indicate that it should not bioaccumulate. VP indicates volatility may occur and inhalation may be a problem. Based on chemical structure and the amine and nitroso moieties, it is predicted to be a carcinogen. In fact, it is a confirmed carcinogen.

#43. N-Nitrosodiphenylamine: mol. wt. 196.23, log Kow 2.79, WS 3.5 X 10^{01} mg/L (UT), & VP 1 X 10^{-01} mm Hg (UT). This chemical is a questionable carcinogen.

Predictive Equation Analysis

$$\log BCF = 0.76 \log Kow - 0.23 \qquad \text{(eq. 18.19)}$$
$$= 0.76\,(2.79) - 0.23$$
$$= 1.8904$$

BCF $= 77.6962$

Kow $= 616.595$

Fate and Transport Analysis

- WS indicates that this chemical could leach, runoff, and be biodegraded, but it could go either way.
- BCF indicates that it should not bioaccumulate.
- Kow indicates that it may bioaccumulate.
- VP indicates that it may be volatile and phototransformation may occur.

Exposure Analysis

WS indicates that this chemical could leach, runoff, and be biodegraded. This is supported by the BCF value, which indicates that it should not bioaccumulate. However, all this is contradicted by the Kow value, which indicates that it could bioaccumulate. Since the WS, BCF, and Kow values are close enough to the bioaccumulation side, it may be better to be on the safe side and predict that this chemical could bioaccumulate and be persistent in the environment. VP indicates that it may be volatile and inhalation may be a problem. Based on chemical structure and the amine and nitroso moieties, it is predicted to be a carcinogen. In fact, it is a confirmed carcinogen.

#44. N-Nitrosodipropylamine: mol. wt. 130.19 g/mol, log Kow 1.31, & WS 9.9 X 10^{03} mg/L at 25° C. This chemical is a confirmed carcinogen.

Predictive Equation Analysis

$$\text{log BCF} = 0.76 \text{ log Kow} - 0.23 \qquad \text{(eq. 18.19)}$$
$$= 0.76 (1.31) - 0.23$$
$$= 0.7656$$

BCF $= 5.8291$

Kow $= 20.4174$

Fate and Transport Analysis

- WS indicates that this chemical could leach, runoff, and be biode-graded.
- BCF and Kow indicate that it should not bioaccumulate.

Exposure Analysis

WS indicates that this chemical could leach, runoff, and be biodegraded. This is supported by the BCF and Kow values, which indicate that bioaccumulation should not occur. Based on chemical structure and the amine and nitroso moieties, it is predicted to be a carcinogen. In fact, it is a confirmed carcinogen.

#45. N-Nitrosomethylethylamine: mol. wt. 88.11 g/mol. This chemical is a confirmed carcinogen.

Predictive Equation Analysis

No data to put into equations.

Fate and Transport Analysis

No data to predict.

Exposure Analysis

Based on the chemical structure, this chemical contains an amine and nitroso moiety; thus, it is predicted to be a carcinogen. In fact, it is a confirmed carcinogen.

#46. N-Nitrosomorpholine: mol. wt. 116.11 g/mol, & WS soluble. This chemical is a confirmed carcinogen.

Predictive Equation Analysis

No data to put into equations.

Fate and Transport Analysis

• WS is stated to be soluble; thus, it could leach, runoff, and be biodegraded, and should not adsorb to soil's organic matter and should not bioaccumulate.

Exposure Analysis

It is predicted that this chemical could leach, runoff, and be biodegraded, should not adsorb to soil's organic matter, and should not bioaccumulate. Based on the nitroso moiety, it is predicted to be a carcinogen. In fact, it is a confirmed carcinogen.

#47. N-Nitrosopiperdine: mol. wt. 114.15 g/mol, & WS soluble. This chemical is a confirmed carcinogen.

Predictive Equation Analysis

No data to put into equations.

Fate and Transport Analysis

• WS is stated to be soluble; thus, it could leach, runoff, and be biodegraded, should not adsorb to soil's organic matter, and should not bioaccumulate.

Exposure Analysis

It is predicted that this chemical could leach, runoff, and be biodegraded, should not adsorb to soil's organic matter, and should not bioaccumulate. Based on the nitroso moiety, it is predicted to be a carcinogen. In fact, it is a confirmed carcinogen.

#48. N-Nitrosopyrrolidine: mol. wt. 100.11 g/mol, & WS soluble. This chemical is a confirmed carcinogen.

Predictive Equation Analysis

No data to put into equations.

Fate and Transport Analysis

- WS is stated to be soluble; thus, it could leach, runoff, and be biodegraded, should not adsorb to soil's organic matter, and should not bioaccumulate.

Exposure Analysis

It is predicted that this chemical could leach, runoff, and be biodegraded, should not adsorb to soil's organic matter, and should not bioaccumulate. Based on the nitroso moiety, it is predicted to be a carcinogen. In fact, it is a confirmed carcinogen.

#49. 5-Nitro-o-toluidine: no data. This chemical is a confirmed carcinogen.

Predictive Equation Analysis

No data to put into equations.

Fate and Transport Analysis

No data to predict.

Exposure Analysis

Based on a chemical structure that contains an amine moiety, it is predicted to be a carcinogen. In fact, it is a confirmed carcinogen.

#50. Pentachloroethane: mol. wt. 202.29 g/mol, log Kow 2.89, WS 5 X 10^{02} mg/L (UT), & VP 3.4 X 10^{00} mm Hg at 20° C. This chemical is a questionable carcinogen.

Predictive Equation Analysis

$$\log BCF = 0.76 \log Kow - 0.23 \qquad \text{(eq. 18.19)}$$
$$= 0.76\,(2.89) - 0.23$$
$$= 1.9664$$
$$BCF = 92.555$$

Kow   = 776.247

$\log$ Koc = $-0.557 \log$ S + 4.277 (μ mol/L)   (eq. 18.16)

$\quad\quad\quad$ = -0.557 (3.393) + 4.277

$\quad\quad\quad$ = 2.387

Koc   = 243.838

Fate and Transport Analysis

• WS indicates that this chemical could leach, runoff, and be biodegraded.
• Koc indicates that it should not adsorb to soil's organic matter.
• BCF indicates that it should not bioaccumulate.
• Kow indicates that bioaccumulation is possible.

Exposure Analysis

WS indicates that this chemical could leach, runoff, and be biodegraded. This is supported by the Koc value, which indicates that it should not be adsorbed to soil's organic matter. It is further supported by the BCF value, which indicates that it should not bioaccumulate. Kow indicates that it may bioaccumulate. This is contradictory to the other predictive data, but useful. BCF and WS data could be predicted either way. Since this chemical is a confirmed carcinogen, it may be better to error on the side of safety. Thus, it is predicted that the chemical may bioaccumulate and exposure may be a problem.

#51. Pentachloronitrobenzene: mol. wt. 296.36 g/mol, & VP 1.35 X 10^{-02} mm Hg at 25° C. This chemical is a questionable carcinogen.

Predictive Equation Analysis

No data to put into the equations.

Fate and Transport Analysis

• VP indicates that volatility and phototransformation may occur.

Exposure Analysis

VP indicates that this chemical may volatilize; thus, inhalation may be a problem. It is predicted to be a carcinogen, based on the nitro moiety. In fact, it is a questionable carcinogen.

#52. Phenacetin or N-(4-ethoxyphenyl) acetamide: mol. wt. 179.21 g/mol, log Kow 1.58, & WS 7.6 X 10^{02} mg/L (UT). This chemical is a confirmed carcinogen.

Predictive Equation Analysis

log BCF= 0.76 log Kow – 0.23   (eq. 18.19)

      = 0.76 (1.58) – 0.23

      = 0.9708

BCF   = 9.3498

Kow   = 38.0189

Fate and Transport Analysis

- WS indicates that this chemical could leach, runoff, and be biodegraded.
- BCF and Kow indicate that it should not bioaccumulate.
- VP indicates that it may volatilize and phototransformation may occur.

Exposure Analysis

WS indicates that this chemical could leach, runoff, and be biodegraded. This is supported by the BCF and Kow values, which indicate that it should not bioaccumulate. VP indicates that it may be volatile and inhalation may be a problem. It is predicted to be a carcinogen, based on the amine moiety. In fact, it is a confirmed carcinogen.

#53. p-Phenylenediamine or 1,4-benzenediamine: mol. wt. 108.14 g/mol, log Kow –0.26, & WS 3.8 X 10^{04} mg/L at 24° C. This chemical is a questionable carcinogen.

Predictive Equation Analysis

log BCF= 0.76 log Kow – 0.23   (eq. 18.19)

      = 0.76 (–0.26) – 0.23

      = –0.4276

BCF   = 0.3736

Kow   = 0.5888

Fate and Transport Analysis

- WS indicates that this chemical could leach, runoff, and be biodegraded.
- BCF and Kow indicate that it should not bioaccumulate.

Exposure Analysis

WS indicates that this chemical could leach, runoff, and be biodegraded. This is supported by the values for BCF and Kow, which indicate that it should not bioaccumulate. It is predicted to be a carcinogen, based on the diamine moiety. In fact, it is a questionable carcinogen.

#54. 2-Picoline or 2-Methyl pyridine: mol. wt. 93.13 g/mol, log Kow 1.06, WS miscible, & VP 1.0×10^{01} mm Hg at 24.4° C.

Predictive Equation Analysis

$$\log BCF = 0.76 \log Kow - 0.23 \qquad \text{(eq. 18.19)}$$
$$= 0.76\,(1.06) - 0.23$$
$$= 0.5756$$

BCF $= 3.7636$

Kow $= 11.4815$

Fate and Transport Analysis

- WS, it is stated to be miscible; thus, it may be soluble.
- BCF and Kow indicate that this chemical should not bioaccumulate.
- VP indicates that it may volatilize and phototransformation may occur.

Exposure Analysis

BCF and Kow indicate that this chemical should not bioaccumulate. WS is stated to be miscible; thus, it is predicted that this chemical could leach, runoff, and be biodegraded. This is predicted by the values for BCF and Kow. VP indicates that it may be volatile and inhalation may be a problem.

#55. Pronamide: mol. wt. 256.13 g/mol, log Kow 3.43, WS 1.5×10^{01} mg/L at 25° C, & VP 8.5×10^{-05} mm Hg (UT). This chemical is a questionable carcinogen.

Predictive Equation Analysis

$$\log BCF = 0.76 \log Kow - 0.23 \qquad \text{(eq. 18.19)}$$
$$= 0.76 (3.43) - 0.23$$
$$= 2.3768$$

BCF $= 238.1223$

Kow $= 2,691.5348$

Fate and Transport Analysis

- WS indicates that this chemical could go either way in regards to leaching, runoff, biodegradation, and soil adsorption.
- BCF and Kow indicate that it should bioaccumulate.
- VP indicates that it may be volatile and phototransformation may occur.

Exposure Analysis

BCF and Kow indicate that this chemical should bioaccumulate. Based on these values, it is predicted that it should not leach, could runoff with soil particles, be adsorbed to soil's organic matter, and should not biodegrade. VP indicates that it may volatilize; thus, inhalation may be a problem. Based on chemical structure, it is predicted to be a carcinogen because of the amide moiety. In fact, it is a questionable carcinogen.

#56. Propionitrile or ethyl cyanide: mol. wt. 55.08 g/mol, log Kow 0.04, WS 1.03×10^{05} mg/L at 25° C, & VP 4.0×10^{01} mm Hg at 22° C.

Predictive Equation Analysis

$$\log BCF = 0.76 \log Kow - 0.23 \qquad \text{(eq. 18.19)}$$
$$= 0.76 (0.04) - 0.23$$
$$= -0.1996$$

BCF $= 0.6315$

Kow $= 1.0965$

Fate and Transport Analysis

- WS indicates that this chemical could leach, runoff, and be biodegraded.
- BCF and Kow indicate that it should not bioaccumulate.
- VP indicates that it may volatilize and phototransformation may occur.

Exposure Analysis

WS indicates that this chemical could leach, runoff, and be biodegraded. This is supported by the values for BCF and Kow, which indicate that it should not bioaccumulate. And these values support the prediction that it should not accumulate in soil and this is supported by the WS value. VP indicates that it may volatilize and inhalation may be a problem. This chemical contains a cyanide moiety, and inhalation may be a problem.

#57. n-Propylbenzene: mol. wt. 120.19 g/mol, log Kow 3.68, WS 6×10^{01} mg/l at 15° C, & VP 2.5×10^{00} mm Hg at 20° C.

Predictive Equation Analysis

log BCF = 0.76 log Kow – 0.23 (eq. 18.19)
  = 0.76 (3.68) – 0.23
  = 2.5668
BCF = 368.8077
Kow = 4,786.3009

Fate and Transport Analysis

- WS indicates that this chemical could leach, runoff, and be biodegraded, but it could go either way.
- BCF and Kow indicate that it should bioaccumulate.
- VP indicates that it may be volatile and phototransformation may occur.

Exposure Analysis

WS indicates that this chemical could leach, runoff, and be biodegraded. This is contradictive in that BCF and Kow indicate that it should bioaccumulate. However, WS indicates it could go either way. Based on the values for BCF and Kow, it is predicted that it should not leach, could runoff with soil particles, could be adsorbed to soil's organic matter, and could accumulate in soil. VP indicates it may be volatile and inhalation may be a problem.

#58. Pyridine: mol. wt. 79.10 g/mol, log Kow 0.65, WS miscible, & VP 1.4 $\times 10^{01}$ mm Hg at 20° C.

Predictive Equation Analysis

$$\log BCF = 0.76 \log Kow - 0.23 \qquad \text{(eq. 18.19)}$$
$$= 0.76 \, (0.65) - 0.23$$
$$= 0.264$$
$$BCF = 1.8365$$
$$Kow = 4.4668$$

Fate and Transport Analysis

- WS states that this chemical is miscible; therefore, it could leach, runoff, and be biodegraded.
- BCF and Kow indicate that it should not bioaccumulate.
- VP indicates that it may volatilize and phototransformation may occur.

Exposure Analysis

WS indicates that this chemical could leach, runoff, and be biodegraded. This is supported by the values for BCF and Kow, which indicate that it should not bioaccumulate. This also supports the prediction that it will not accumulate or be absorbed to soil's organic matter. VP indicates volatility and inhalation may be a problem.

#59. Safrole or 5-(2-propenyl)-1,3-benzodioxole: mol. wt. 162.19 g/mol, & WS insoluble. This chemical is a confirmed carcinogen.

Predictive Equation Analysis

No data to put into equations.

Fate and Transport Analysis

- WS states that it is insoluble; therefore, it should not leach, could runoff with soil particles, and should adsorb to soil's organic matter.

Exposure Analysis

WS indicates that this chemical it insoluble; therefore, it is predicted that it should not leach or biodegrade, could runoff with soil particles, should accumulate in soil's organic matter, and should bioaccumulate. Safrole moiety is predicted to be a carcinogen and it is, in fact, a confirmed carcinogen.

#60. Styrene or ethenylbenzene: mol. wt. 104.15 g/ mol, log Kow 3.16, WS 3 X 10^{02} mg/L at 20° C, & VP 5 X 10^{00} mm Hg at 20° C. This chemical is a questionable carcinogen.

Predictive Equation Analysis

log BCF = 0.76 log Kow – 0.23   (eq. 18.19)
        = 0.76 (3.16) – 0.23
        = 2.1716
BCF   = 148.4568
Kow   = 1,445.4398

Fate and Transport Analysis

- WS indicates that this chemical could leach, runoff, and be biodegraded, but it could go either way.
- BCF and Kow indicate that it should bioaccumulate.
- VP indicates that it may be volatile and phototransformation may occur.

Exposure Analysis

WS indicates that this chemical could leach, runoff, and be biodegraded. However, BCF and Kow indicate that it should bioaccumulate. The WS value means that it could go either way. To be on the safe side, it is predicted that this chemical should not leach, could runoff with soil particles, and should adsorb to soil's organic matter, and this is supported by the values for BCF and Kow. VP indicates that volatilization may occur and inhalation may be a problem, as it is a questionable carcinogen.

#61. 2,3,7,8-Tetrachlorodibenzo-p-dioxin or 2,3,7,8-TCDD: mol. wt. 321.96 g/mol, log Kow 6.64, WS 1.93 X 10^{-05} mg/L (UT), & VP 7.4 X 10^{-10} mm Hg at 25° C. This chemical is a confirmed carcinogen.

Predictive Equation Analysis

log BCF = 0.76 log Kow – 0.23   (eq. 18.19)
        = 0.76 (6.64) – 0.23
        = 4.8164
BCF   = 65,523.9394
Kow   = 4,365,158.322

Fate and Transport Analysis

- WS indicates that this chemical should not leach, could runoff with soil particles, and should be adsorbed and accumulated in soil.
- BCF and Kow indicate that it should be bioaccumulated.
- VP indicates that it should not be volatile.

Exposure Analysis

WS indicates that this chemical should not leach, could runoff with soil's organic matter, should accumulate in soil, and should adsorb to soil's organic matter. This is supported by the values for BCF and Kow, which indicate that it should bioaccumulate. VP indicates that it should not be volatile. Based on chemical structure, a dibenzo and a dioxin, it is predicted to be a carcinogen. In fact, it is a confirmed carcinogen.

#62. 1,1,1,2-Tetrachloroethane: mol. wt. 167.85 g/mol, log Kow 3.04, WS 2×10^{02} mg/L at 20° C, & VP 1.0×10^{01} mm Hg at 19.3° C.

Predictive Equation Analysis

$$\log BCF = 0.76 \log Kow - 0.23 \qquad \text{(eq. 18.19)}$$
$$= 0.76\,(3.04) - 0.23$$
$$= 2.84$$

BCF   = 120.337

Kow   = 1,096.48

Fate and Transport Analysis

- WS indicates that this chemical could go either way. BCF and Kow indicate that it should bioaccumulate.
- These values support that it should not leach, could runoff with soil particles, should not biodegrade, and should accumulate in soil's organic matter.
- VP indicates that it may volatilize and phototransformation may occur.

Exposure Analysis

BCF and Kow indicate that this chemical should bioaccumulate. Based on these values, it is predicted that it should not leach, could runoff with soil particles, and could be adsorbed to soil's organic matter. VP indicates that it may be volatile and inhalation may be a problem, as it is a questionable carcinogen.

#63. 2,3,4,6-Tetrachlorophenol: mol. wt. 231.89 g/mol, & VP 6.0 X 10^{01} mm Hg at 190° C.

Predictive Equation Analysis

No data to put into equations.

Fate and Transport Analysis

• VP indicates that this chemical should not volatilize at standard conditions and temperatures.

Exposure Analysis

Cannot be assessed.

#64. Tetraethyl dithiopyrophosphate or Sulfotep; mol. wt. 322.31 g/mol, log Kow 3.83, WS 2.5 X 10^{01} mg/L (UT), & VP 1.7 X 10^{-04} mm Hg at 20° C.

Predictive Equation Analysis

$$\log BCF = 0.76 \log Kow - 0.23 \qquad \text{(eq. 18.19)}$$
$$= 0.76 (3.83) - 0.23$$
$$= 2.6808$$

BCF $= 479.5126$

Kow $= 6,760.8298$

Fate and Transport Analysis

• WS indicates that this chemical could go either way.
• BCF and Kow indicate that it should bioaccumulate.
• VP indicates that it may be volatile and phototransformation may occur.

Exposure Analysis

BCF and Kow indicate that this chemical should bioaccumulate. Based on these values and the WS, which indicates that it could go either way, it is predicted that it should not leach, but it could runoff with soil particles, adsorb to soil's organic matter, and accumulate in soil. VP indicates that volatility is possible and inhalation may be a problem.

#65. 1,2,3-Trichlorobenzene: mol. wt. 181.45 g/mol, log Kow 4.14, & WS 1.2 X 10^{01} mg/L at 22° C.

Predictive Equation Analysis

$$\log \text{BCF} = 0.76 \log \text{Kow} - 0.23 \qquad \text{(eq. 18.19)}$$
$$= 0.76 (4.14) - 0.23$$
$$= 2.9164$$

BCF   = 824.8975

Kow   = 13,803.8426

Fate and Transport Analysis

- WS indicates that it could go either way.
- BCF and Kow indicate that it should bioaccumulate.

Exposure Analysis

BCF and Kow indicate that this chemical should bioaccumulate. Based on the values for BCF and Kow, it is predicted that it should not leach, could runoff with soil particles, and could be adsorbed into the soil's organic matter. This is supported by the low WS value, which indicates that it could go either way.

#66. 1,1,1-Trichloroethane or methylchloroform: mol. wt. 133.40 g/mol, log Kow 2.47, WS 4.4 X 10^{03} mg/L at 20° C, & VP 1.0 X 10^{02} mm Hg at 20° C. This chemical is a questionable carcinogen.

Predictive Equation Analysis

$$\log \text{BCF} = 0.76 \log \text{Kow} - 0.23 \qquad \text{(eq. 18.19)}$$
$$= 0.76 (2.47) - 0.23$$
$$= 1.6472$$

BCF   = 44.3813

Kow   = 295.1209

Fate and Transport Analysis

- WS indicates that this chemical could leach, runoff, and be biodegraded.
- BCF and Kow indicate that it should not bioaccumulate.
- VP indicates that it may volatilize and phototransformation may occur.

Exposure Analysis

WS indicates that this chemical could leach, runoff, and be biodegraded. It also indicates that it should not accumulate or adsorb in soil's organic matter. This is supported by BCF and Kow values, which indicate that it should not bioaccumulate. VP indicates that it may volatilize and inhalation may be a problem, as it is a questionable carcinogen.

#67. 1,1,2-Trichloroethane: mol. wt. 133.40 g/mol, log Kow 2.17, WS 4.5 X 10^{03} mg/L at 20° C, & VP 2.5 X 10^{01} mm Hg at 25° C. This chemical is a suspected carcinogen.

Predictive Equation Analysis

$$\log BCF = 0.76 \log Kow - 0.23 \qquad \text{(eq. 18.19)}$$
$$= 0.76 (2.17) - 0.23$$
$$= 1.4192$$

BCF $= 26.2543$

Kow $= 147.9108$

Fate and Transport Analysis

- WS indicates that this chemical could leach, runoff, and be biodegraded.
- BCF and Kow indicate that it should not bioaccumulate.
- VP indicates that it may volatilize and phototransformation may occur.

Exposure Analysis

WS indicates that this chemical could leach, runoff, and be biodegraded. It also indicates that it should not accumulate or be adsorbed in soil's organic matter. This is supported by BCF and Kow values, which indicate that it should not bioaccumulate. VP indicates that it may volatilize and inhalation may be a problem, as it is a suspected carcinogen.

#68. Trichloroethylene or trichloroethene: mol. wt. 131.39 g/mol, log Kow 2.53, WS 1.1 X 10^{03} mg/L at 25° C, & VP 7.7 X 10^{01} mm Hg at 25° C. This chemical is a suspected carcinogen.

Predictive Equation Analysis

$$\log BCF = 0.76 \log Kow - 0.23 \qquad \text{(eq. 18.19)}$$
$$= 0.76\,(2.53) - 0.23$$
$$= 1.6928$$

BCF $= 49.2947$

Kow $= 338.8442$

Fate and Transport Analysis

- WS indicates that this chemical could leach, runoff, and be biodegraded.
- BCF and Kow indicate that it should not bioaccumulate.
- VP indicates that it may volatilize and phototransformation may occur.

Exposure Analysis

WS indicates that this chemical could leach, runoff, and be biodegraded. It also indicates that it should not accumulate or be adsorbed in soil's organic matter. This is supported by BCF and Kow values, which indicate that it should not bioaccumulate. VP indicates that it may volatilize and inhalation may be a problem, as it is a suspected carcinogen.

#69. 1,2,3-Trichloropropane: mol. wt. 147.43 g/mol, WS is insoluble & VP 2 X 10^{00} mm Hg at 20° C.

Predictive Equation Analysis

No data to put into equations.

Fate and Transport Analysis

- WS is stated to be insoluble; thus, this chemical should not leach, could runoff with soil particles, and should not be biodegraded.
- VP indicates that it may by volatile and phototransformation may occur.

Exposure Analysis

WS indicates that it should not leach, could runoff, and should not be biodegraded. Based on it being insoluble, it is predicted that it could accumulate in soil's organic matter and bioaccumulate. VP indicates that it may volatilize and inhalation may be a problem.

#70. 1,2,4-Trimethylbenzene: mol. wt. 120.19 g/mol, log Kow 3.63, WS 5.7 X 10^{01} mg/L at 20° C, & VP 1 X 10^{00} mm Hg at 13.6° C.

Predictive Equation Analysis

$$\log BCF = 0.76 \log Kow - 0.23 \qquad \text{(eq. 18.19)}$$
$$= 0.76 (3.63) - 0.23$$
$$= 2.5288$$

BCF $= 337.9092$

Kow $= 4,265.7952$

Fate and Transport Analysis

• WS indicates that this chemical could go either way.
• BCF and Kow indicate that it should bioaccumulate.
• VP indicates that it may be volatile and phototransformation may occur.

Exposure Analysis

BCF and Kow indicate that this chemical should bioaccumulate. Based on the values for BCF and Kow, it is predicted that it should not leach, could runoff with soil's organic matter, should accumulate in soil, and could be adsorbed in soil's organic matter. VP indicates that it may volatilize and inhalation may be a problem.

#71. 1,3,5-Trimethylbenzene: mol. wt. 120.19 g/mol, Kow 3.42, WS 2 X 10^{01} mg/L (UT), & VP 1 X 10^{00} mm Hg at 9.5° C.

Predictive Equation Analysis

$$\log BCF = 0.76 \log Kow - 0.23 \qquad \text{(eq. 18-19)}$$
$$= 0.76 (3.42) - 0.23$$
$$= 2.3692$$

BCF $= 233.9915$

Kow   = 2,630.268

Fate and Transport Analysis

- WS indicates that this chemical could go either way.
- BCF and Kow indicate that it should bioaccumulate.
- VP indicates that it may be volatile and phototransformation may occur.

Exposure Analysis

BCF and Kow indicate that this chemical should bioaccumulate. Based on the values for BCf and Kow, it is predicted that it should not leach, could runoff with soil's organic matter, should accumulate in soil, and could be adsorbed in soil's organic matter. VP indicates that it may volatilize and inhalation may be problem.

#72. Vinyl acetate: mol. wt. 86.09 g/mol, log Kow 0.73, WS 2×10^{04} mg/L (UT), & VP 1×10^{02} mm Hg at 23.3° C. This chemical is a questionable carcinogen.

Predictive Equation Analysis

$$\log BCF = 0.76 \log Kow - 0.23 \qquad \text{(eq. 18.19)}$$
$$= 0.76 (0.73) - 0.23$$
$$= 0.3248$$

BCF   = 2.1125
Kow   = 5.3703

Fate and Transport Analysis

- WS indicates that this chemical could leach, runoff, and be biodegraded.
- BCF and Kow indicate that it should not bioaccumulate.
- VP indicates that it may volatilize and phototransformation may occur.

Exposure Analysis

WS indicates that this chemical could leach, runoff, and be biodegraded. It is predicted that it should not adsorb to soil's organic matter. This is supported by the values for BCF and Kow, which indicate that it should not bioaccumulate. VP indicates that it may be volatile and inhalation may be a problem. It is predicted to be a carcinogen, based on the vinyl moiety. In fact, it is a questionable carcinogen.

#73. Xylene including ortho, meta and para: mol. wt. 106.17 g/mol, log Kow 2.77 to 3.20, WS insoluble, & VP 1.0 X 10^{01} mm Hg at 27.3 to 32.1° C.

Predictive Equation Analysis

$$\log BCF = 0.76 \log Kow - 0.23 \qquad \text{(eq. 18.19)}$$
$$= 0.76\,(2.77) - 0.23$$
$$= 1.8752$$
BCF $= 75.024$

Kow $= 588.8437$

$$\log BCF = 0.76 \log Kow - 0.23 \qquad \text{(eq. 18.19)}$$
$$= 0.76\,(3.20) - 0.23$$
$$= 2.202$$
BCF $= 159.2209$

Kow $= 1,584.8932$

Fate and Transport Analysis

• WS is stated to be insoluble; thus, this chemical should not leach, could runoff with soil particles, should not biodegrade, and should accumulate in soil's organic matter.
• Koc indicates that it should not adsorb to soil's organic matter, and this contradicts other WS, Kow and BCF data.
• Kow indicates that it is lipo soluble and should bioaccumulate.
• BCF indicates that it should bioaccumulate.

Exposure Analysis

WS indicates that this chemical should not leach, could runoff, should not be biodegraded, and should accumulate in soil's organic matter. This is supported by the Kow and BCF values, which indicate that it is lipo soluble and should bioaccumulate. However, the Koc value indicates that it should not accumulate and not persist in soil. Since WS, BCF, and Kow support bioaccumulation, it is predicted to bioaccumulate and to be persistent in the environment. VP indicates that it may volatilize and inhalation may be a problem.

REFERENCES

[1]   Layman, Warren J., William F. Reehl, and David H. Rosenblatt, Ph.D., *Handbook of Chemical Property Estimation Methods*, McGraw-Hill Book Company, 1982.
[2]   EPA *Handbook of RCRA Ground-Water Monitoring Constituents: Chemical & Physical Properties*, Office of Solid Waste, September 1992.
[3]   Lewis, J. Richard, Jr., *Hazardous Chemicals Desk Reference, Second Edition,* Van Nostrand Reinhold, 1991.

Glossary

Absorbed/absorption. The ability of a chemical to be sorbed to soil and be released from soil with water or other liquid solvents.

Adsorbed/adsorption. The ability for a chemical to sorb to soil and not be released with water and most other liquid solvents.

Active ingredients. An agent (i.e., a chemical, biological, etc.) that makes a product do what it says it will do, a pesticide.

Aggregate effects. The toxic effects, added effects, that all the chemicals to which one is exposed would have on humans or the environment.

Antagonistic effect. Different reactions caused by opposite chemical forces.

Antibacterial. An ingredient used as an antiseptic to kill or control microbes/bacteria.

Asphyxia. Death cause by lack of oxygen.

Biometabolism. The precess by which microbes break down or transform a chemical to a new chemical(s); also called biodegradation.

Break-down products. Products formed from the breaking down of a chemical. These products may or may not be more toxic than parent chemical, the starting material.

Carcinogen. Causes cancer.

Catalase. An enzyme in blood and tissue that contributes to the decomposition of hydrogen peroxide to form water and oxygen.

Coal tar pitch. Is what is left after coal tar (from coal) is redistilled.

Co-biometabolism. A substance that would not ordinarily be biometabolized/biodegraded without an outside influence.

Co-metabolism. A substance that would not ordinarily be metabolized without an outside influence.

Contamination/ed. Air, water, soil, plants, animals and humans containing a chemical, microbe, virus, etc. that is not naturally occurring in that environment.

Edema. Excess accumulation of fluids, accumulation of water, can cause swelling.

Electromagnetic forces (EMF). The radiation given off by electric forces/ fields.

Fetotoxin. Toxic effects to the fetus, can kill the fetus.

Glottis. Part of the larynx.

Hazardous. A toxin or toxic chemical that is harmful to plants or animals, including humans.

Hydrocarbon. An organic chemical consisting only of the elements carbon and hydrogen.

Inert ingredients. An ingredient in a product that takes no part in how the product performs, not an active ingredient.

Leaching. The ability of a chemical to move through soil with water or some other liquid solvent.

Metabolism. The process by which plant or animals break down or transform chemicals to another chemical(s).

Mutagen. Causes mutation in living organisms.

Neurotoxin. Causes brain damage.

Oncogen. Causes tumors.

Pesticide. An active ingredient used to mitigate a pest (i.e., weeds, insects, fish, rodents, microbes, etc.).

Photo-degradation/transformation. The ability for a chemical to absorb light (sunlight) and break-down or build-up (to lower or heavier molecular weight molecules.

Residues. A pesticide residue is only the parent pesticide; however, pesticide residues include parent chemical and its break-down chemicals.

Risk assessment. A process to determine the potential risk when exposed to a toxin.

Runoff. The ability for a chemical to move across soil with water or some other liquid solvent, or with soil particles.

Sublime. The passage of a chemical from one physical state to another, form a solid to a gas. The liquid state is skipped.

Synergistic. The effects that one or more chemical have on other chemicals when brought together.

Teratogen. Causes birth defects/abnormalities from conception and birth.

Transformation. The break down of a chemical to form a new chemical of larger or smaller molecular structure.

Vapor. A chemical in its gaseous (volatile) state.

Volatile/volatilization. The ability for a chemical to become a vapor and be released in the air.

Wood preservative. A pesticide that is used to protect wood from insects or microbes.

Index

AAF, 256
Absorption, 3, 291
Accumulation, 4
Acenaphthene, 32, 289, 291
Acenaphthylene. 32-33, 289
Acephate, 33
Acetamide, 257
Acetic acid amine, 257
Acetone, 289, 292
Acetonitrile, 289, 292
Acetophenone, 34
2-Acetylaminofluorene, 256, 311
Acid rain, 238, 285
Acridine, 34-35
Acrolein, 255, 289, 293
Acrylamide, 257
Acryaldehyde, 255
Acrylonitrile, 35-36
Adsorption, 3, 19, 181-82, 292
Adverse effects, 176
Agent orange, 282
Air, 174
Air compartments, 5, 16
Air contamination, 5
Alachlor, 36, 256
Alar, 262
Aldehyde, 255, 263
Aldicarb, 37
Aldrin, 38, 281
Allyl chloride, 312
Amiben, 58
Amine, 256, 263
Amino, 256, 263
4-Aminobiphenyl, 289, 294
Amide, 256-57, 263
Amido, 261, 263
para-Aminoazobenzen hydrochloride, 257
para-Aminobenzene, 256

para-Aminobenzene hydrochloride, 256
4-Aminodiphenyl, 259
ortho-Aminobiphenyl, 259
para-Aminobiphenyl, 259
para-Aminophenol, 256
Amitraz, 257
Amitrole, 38-39
Ammonia, 39
Anilide, 256, 263
Aniline, 39-40, 257, 263, 313
Animals, 174
Animal compartments, 7, 16
Anthon, 165-66
Anthracene, 40-41
Aquatic environment, 174
Aroclors, 148, 282
Asbestos, 41-42, 283
Asulan, 260
Atrazine, 42-43, 281
Attenuation, 183
Avadex, 80-81, 260
Azo, 258, 263

Banvel D, 83-84
Basagran, 43-44
Basalin, 109, 258
Bay 9010, 150
Baygon, 150, 281
BCF. See Bioconcentration, 31, 205, 227
Bendicarb, 281
Benefin, 43
Benomyl, 209, 212, 258
Bentazon, 43-44
Benzenamine, 39-40, 257, 263, 313
Benzene, 44-45, 258, 263
Benzene alcohol, 295
1,4-Benzenediamine, 346

1,2-Benzenedicarboxlic acid, dioctyl ester, 289, 294
Benzenemethanol, 289, 295
Benzidine, 45, 259
Benzo [ ] anthracene, 45
Benzo [ß] fluoranthene, 46
Benzo [k] fluoranthene, 289, 296
Benzo [ghi] perylene, 290, 297
Benzo [ ] pyrene, 47
1,2-Benzphenanthrene, 72, 265
Binfenox, 47
2,3:4,5-Bis(2-butlene0   tetrahydro-2-furaldehyde, 255
Bioaccumulation, 2, 4, 26-27, 227
Bioconcentraton, 31, 227
Bioconcentration factor(BCF), 31
Bioconcentration   factor   flow-through((BFC)f), 228
Bioconcentration factor static((BCF)t), 228
Biodegradation, 2, 25-26, 73
Bioxone, 125
Biphenyl, 48, 259, 263
1,1'-Biphenyl-4-amine, 289, 294
Bladex, 73-74
Botran, 258
Bound, 183
Breakdown products, 291
Bromacil, 49, 265
Bromobenzene, 49-50, 290, 297
Bromodichloromethane, 50
Bromoform, 51
Bromomethane, 52-53
4-Bromophenyl phenyl ether, 52-53
Bufencarb, 53
Butachlor, 256
Butam, 265
Butralin, 53-54
Buturon, 262
Butylbenzene, 290, 298
Butyl benzyl phthalate, 54
x-sec- Butyl-4-chlorodiphenyloxide, 54-55
N-sec-Butyl-4-tert-butyl-2,6-dinitroaniline, 258
Bux, 53
1,1'-Biphenyl-4-amine

Calcium ethylenebisdithiocarbamate, 265
Captafol, 260
Captan, 55-56, 197, 199, 260, 261, 281
Carbaryl, 56, 187, 190, 197-98, 281
Carbofuran, 57

Carbolic acid, 144-45
Carbon tetrachloride, 57-58
Carboxin, 257
Carzol, 111
Casoron, 84-85
Cation exchange capacity(CEC), 20
CEC, 20
CDEC, 260
Chemicals, movement of, in environment, 2-3
Chemical breakdown, mechanisms of, 1-2
Chemical mixture, 178
Chloramben, 58
Chlorambucil, 256
Chlorbromuron, 59
Chloroaniline, 255, 261, 263
p-Chloroaniline, 290, 300, 314
Bis(2-Chloro-1-methylether
Chlordane, 59-60, 232, 281
Chlordimeform, 257
Chlorine, 60
Chlormethazole, 122-23
2-Chloroallyl diethyldithiocarbamate, 260
4-Chlorobenzamine, 290, 300
Chlorobezene, 61
Chlorobenzilate, 290, 299
4-Chlorobiphenyl, 61-62
Chlorobromuron, 262
2-Chloro-1,3-butadiene, 316
Chlorodiphenyl, 259
4-Chlorodiphenyl oxide, 63
Chloroethane, 63-64
Bis (2-Chloroethoxy) methane, 64-65
Bis (2-Chloroethyl) ether, 65
Chloroethene, 171
Chloroform, 65-66
Chloromethane, 66-67
p-Chloro-m-cresol, 62-63
1-Chloro-2-methylbenzene, 290, 303
Bis (2-Chloromethyl) ether, 67-68
Bia (2-Chloeo-1-methylether) ether, 315
2-Chloronaphthalene, 290, 301
Chloroneb, 68
2-Chlorophenol, 68-69
o-Chlorophenol, 68-69
4-Chlorophenyl phenyl ether, 69, 290, 303
3-Chloro-1-propene, 312
Chloroprene, 316
Chlorothalonil, 281
N-[[(4-Chlorophenyl)   thio]   methyl] phthalimide, 265

2-Chlorotoluene, 290, 303
2-Chloro-1-(2,2-trichlorophenyl) vinyl dimethyl phosphate, 266
2-Chloro-N-[[(4-trifuoromethoxy) phenyl amino] carbonyl] benzamide, 266
Chloroxuron, 70
Chlorphos, 165-66
Chlorpropham, 70-71
Chlorpyrifos, 103-4, 281
Chlorthiamid, 71-72
Chrysene, 72, 265
CIPC, 70-71
Cobex, 96
Cortisol, 264
Cortisone, 264
Counter, 157-58
m-Cresol, 317
Crufomate, 72-73
Cumene, 328
Cyanazine, 73-74
Cycloate, 74
Cygon, 93-94

2,4-D, 75, 187, 192, 281-82
Daconil, 281
Dacthal, 281
Dalapon, 76
Daminozide, 262
DATC, 80-81
DBCP, 76-77, 197, 200, 281, 184
DCPA, 285
DDD, 77-78, 259, 281
DDE, 78, 281
DDT, 11, 14, 79, 259, 281
DDVP, 90, 281
Degradates, 177
Degradation, 173, 177
DEHP, 92
Dehydrostilbestrol, 264
Demosan, 68
DEP, 93
DES, 264
Desorption, 182-83
Dialifor, 79-80
Dialifos, 79-80
Diallate, 80-81, 260, 316
Diamidofos, 81
Diazinon, 82, 209, 211, 231-32, 281
Dibenzofuran, 318
Dibromochloromethane, 82-83

1,2-Dibromo-3-chloropropane, 76-77, 197, 200, 281, 285
Di-n-butyl phthalate, 319
S-(2,3-Dichloroally) diisopropylthio-carbamate, 80-81, 260, 317
1,2-Dichlorobenzene, 85-86
ortho-Dichlorobenzene, 85-86
para-Dichlorobenzene, 86-87, 258
2,3-Dichlorobinzidine
3,3'-Dichlorobenzidine, 259
4,4'-Dichlorobenzilate, 290, 299
4,4'-Dichlorobiphenyl, 87
Dichloro diphenyl dichloroethane (DDD), 77-78, 259
Dichloro diphenyl trichloroethane (DDT), 11, 79, 259
1,2-Dichloroethane, 108
Dichloroethylether, 290, 340
cis-1,3-Dichloropene, 321
2,4-Dichlorophenol, 88-89
2,6-Dichlorophenol, 320
2,4-Dichlorophenoxyacetic acid, 75, 282
1,2-Dichloropropane, 321
2,2-Dichloropropionic acid, 76
2,5-Dichlorothiobenzamide, 71-72
2,2-Dichlorovinyl dimethyl phosphate, 90
Dichlozoline, 257
Dicamba, 83-84
Dichlobenil, 84-85
Dichlofenthion, 85
Dichloromethane, 127-28
3,6-Dichloropicolinic acid, 89-90
Dichlorodifluoromethane, 88
3,6-Dichloropicloinic acid, 89-90
Dichlorvos, 90, 209, 281
Dieldrin, 90-91, 231, 237
Diethylaniline, 91-92
Diethyleneimide oxide, 260
Di-2-ethylhexyl phthalate, 92
Diethyl phthalate, 93
Diethylstilestrol (DES), 264
Diethylstilbestrol dipalmitate, 264
Diethylstilbestrol dipropionate, 264
Diflubenzuron, 95-96, 187, 262, 266
Dimethoate, 93-94
p-(Dimethylamino) azobenzene, 322
4-Dimethylaminobenzene, 256
1,11-Dimethylchrysene, 265
1,2-Dimethyl-5-nitroimidazole, 266
Dimethylnitrosamine, 94
2,4-Dimethylphenol, 323

Dimethyl phthalate, 95
N,N-dimethyl-N'-2-pyridinyl-N'-(2-thienylmethyl)-1,2-ethanediamine, 330
Dimilin, 95-96, 187, 220-21, 262, 266
Dinitramine, 96-97
1,3-Dinitrobenzene, 324
4,6-Dinitro-o-cresol, 97
2,4-Dinitrophenol, 97-98
2,4-Dinitrotoluene, 98
Di-n-octylphthalate, 289, 294, 324
Dinoseb, 98-99
1,4-Dioxane, 290, 305
Dioxin, 281, 284
Di-n-propylnitrosamine, 100-01
Diphenyl, 48, 259, 263
Diphenylamine, 266, 290, 306
Diphenylnitrosamine, 99
Diphenyl oxide, 100
Disodium methanearsonate, 102-03
Dissipation
  in soil, 267
  rate of, 2
Dissociation, 2
Disulfoton, 101
Di-Syston, 101
Dithiodemeton, 101
Diuron, 102, 264
DMP, 95
DNBP, 98-99
Dowco 169, 81
DSMA, 102-03
Dursban, 103-04, 281
Dyanap, 98-99
Dylox, 165-66

EDB, 107-08, 285
Endosulfan I, 290
Endothall, 104
Endrin, 105, 231, 233
Endrin aldehyde, 325
Environmental compartments, 4
    compartment air, 5, 16
    compartment animals, 7, 16
    compartment plants, 6-7, 16
    compartment soil, 5-6, 15-16
    compartment water, 5, 15
    rate of chemical dissipation in , 2
Eptam, 105-06
EPTC, 105-06
Estimated exposure to fish, 271

Estradiol, 264
Estradiol-3-benzoate, 264
Estradiol dipropionate, 264
Estradiol mustard, 264
Estradiol polyester, 264
Estriol, 264
Estrone, 264
Ethenylbenzene, 351
Ethion, 106
Ethoxyclor, 259
N-(4-ethoxyphenyl) acetamide, 346
Ethylbenzene, 107
Ethyl chloride, 63-64
S-Ethyl cyclohexylethylthiocarbamate, 74-75
Ethyl cyanide, 348
S-Ethyl dipropylthiocarbamate, 105-06
Ethylene dibromide, 107-08
Ethylene dichloride, 108
Ethylenethiourea, 259
Ethylene trichloride, 156-57
Ethylenimine, 260
Bis(2-Ethylhexyl) phthalate, 326
ETU, 259
Eurex, 74
Evital, 139-40
Exposure assessment, 29-30
Exposure, 175

Fate and transport, 1-4
Fenchlorphos, 153-54
Fenuron, 108-09
Ferbam, 260
Finuron, 264
Fluchloralin, 109
Fluoranthene, 110
Fluorene, 110-11
N-9H-Fluoren-2-yl-acetamide, 256, 311
Flumeturon, 262
Fluorocarbon-11, 164-65
Fluorocarbon-12, 88
Folpet, 261, 281
Food-chain contamination, 3-4
Formaldehyde, 255, 283
Formetanate, 111
Freon-11, 164-65
Freon-12, 88
Fumazone, 76-77
Furandan, 57

Gardona, 266

Garlon, 166-67
Glyphosate, 111-12
Gophacide, 265

Half-life, 2, 267
Hazardous organics, 179
Heptachlor, 112-13, 231, 238, 281
Hexachlorobenzene, 113, 281
Hexachlorobutadiene, 114
Hexachlorocyclopentadiene, 114-15
Hexachloroethane, 115
Hexachlorophene, 282, 326
2-Hexanone, 290, 307
Higher molecular weight compounds, 181
HPLC, 185
Hydrazine, 262-63
Hydrazine sulfate, 263
Hydrolysis, 2, 15-16, 173, 183
      air compartment, 16
      plant and animal compartments,
      16
      soil compartments, 15
      water compartments, 15

Imidan, 116, 261, 281
Imidazole, 258, 263
Imide, 260, 263
Imine, 260, 263
Immobilization, 180
Indeno(1,2,3-cd)pyrene, 327
Indole, 260, 263
Indigo, 261
Iodomethane. 333
Ipazine, 116-17
IPC, 149-50
Isatin, 261
Isobutyl alcohol, 290, 308
Isocil, 117
Isophorone, 328
Isopropylbenzene, 328
p-Isopropyltoluene, 329
Isosafrole, 265, 330
Isosafrole-n-octylsulfoxide, 265
Isopropalin, 118
o-Isopropoxyphenyl methylcarbamate, 150
Isopropyl carbanilate, 149-50
3-(4-Isopropylphenyl)-1,1-dimethylurea,
   266

Kd, 20, 218
Kelthane, 259

Kepone, 118-19, 231, 235
Koc, 13, 20-21, 205, 217-18
Kom, 20
Korlan, 153-54
Kow, 12-13, 183, 195, 205, 217

Lasso, 36
Leaching, 2, 22, 183, 292
Lead, 119, 283
Leptophos, 120, 231, 239, 282
Lindane, 120-21, 220, 222, 282, 290, 309
Linuron, 121-23, 262

Malathion, 122, 281
Maleic hydrazide(MH), 263
Maneb, 209, 213, 259
MBC, 165, 188, 258
MBK, 332
Metabolism, 2, 292
Metaldehyde, 266
Metalkamate, 53
Methapyrilene, 330
Methazole, 122-23
Methbenzthiazuron, 258
Methomyl, 123-24, 187, 189
Methoxychlor, 124, 189, 191
Methoxy-2-methylpropane
Methyl bromide, 51-52
Methyl butyl ketone, 290, 307
Methyl chloride, 66-67
3-Methylcholanthrene, 127, 331
Methyl chloroform, 126-27, 355
3-Methylchrysene, 265
4-Methylchrysene, 265
5-Methylchrysene, 265
Methylethylketone, 332
Methyl hydrazine, 263
Methyl iodine, 333
Methyl isobutylketone, 335
Methyl methacrylate, 333
Methyl methanesulfonate, 334
2-Methyl-1-propanol, 290, 308
2-Methyl pyridine, 347
Methyl tert-butyl ether, 335
2-Methoxy-2-methylpropane, 335
2-Methoxy-3,5,6-trichloropyridine, 125
9-Methylanthracene, 125-26, 197, 201
Methylene chloride, 127-28
Methyl isothiocyanate, 128-29
2-Methylnaphthalene, 129
Methylparathion, 129-30

3-Methylphenol, 317
Metobromuron, 130-31
Metolachlor, 131, 259
Mexacarbate, 132
mg/L, 11
Microbibal degradation, 173, 183
Mirex, 132-33, 231, 240, 282
Mobility, 178
Modown, 47
Monolinuron, 133, 262
Monometflurazon, 139-40
Monuron, 134, 262

Nabam, 260
Naphthacene, 158
Naphthalene, 134-35
1,4-Naphthaquinone, 336
1-Naphthylamine, 337
2-Naphthylamine, 290, 309, 337
beta-Naphthylamine, 256
1-Naphthol, 135
Neburon, 262
Nellite, 81
Nemacide, 85
Nemagon, 76-77
Neoprene, 316
Nitralin, 136
Nitrapyrin, 136-37

2-Nitroaniline, 338
Nitrobenzene, 137-38
4-Nitrobiphenyl, 259
2-Nitrophenol, 138
4-Nitrophenol, 138-39
4-Nitroquinoline-1-oxide, 339
Nitrosoaniline, 258, 264
N-Nitrosobutylamine, 339
N-Nitrosodiethylamine, 339
N-Nitrosodimethylamine, 340
N-Nitrosodiphenylamine, 341
N-Nitrosodipropylamine, 342
N-Nitrosomethylethylamine, 342
N-Nitrosomorpholine, 343
N-Nitrosopiperdine, 343
N-Nitrosopyrrolidine, 344
5-Nitro-o-toluidine, 344
Nonoccupational Pesticide Exposure Study,
    279, 280
NOPES, 279, 280
Norflurazon, 139-40
N-Serve, 136-37

OC, 20, 217
Octanol water, 12-14, 183
Octanol water partation coefficient (Kow),
    12, 195
OM, 217
Organic carbon, 20, 217
Organic matter, 217
Orthocide, 55-56
Oxadiazon, 140
1,1'-Oxybis(2-chloroethane), 290, 304
Oxychlordane, 281

Paranaphthalene, 40-41
Paraquat, 140-41
Parathion, 141-42, 220
Prtition coefficient (P), 12
Patoran, 130-31
Payze, 73-74
PCBs, 148
PCP, 143, 197, 202, 282
Pebulate, 161-62
Pentachlorobenzene, 142
Pentachloroethane, 344
Pentachloronitrobenzene, 345
Pentachlorophenol, 143, 197, 202, 282
Perchloroethylene, 160
Phenacetin, 346
Phenanthrene, 143-44
Phenol, 144-45, 258
p-Phenylenediamine, 346
Phenyl ether, 100
Phenylhydrazine, 263
ortho-Phenylphenol, 282
Phorate, 146-47
Phosalone, 145
Phosmet, 116
Photodegradation, 17
Photolysis, 1-2, 17
Photooxidation, 17
Phototransformation, 17
Phthalic anhydride, 146
Phthalimide, 261, 263
Phthalthrin, 281
Picloram, 147, 234, 241
2-Picoline, 347
Planavin, 136
Plant compartments, 6-7, 16
Plant comtamination, 6-7
Plants, 154
Poisoning yourself, 277

Poisoning oursely, 277
Polychlorinated biphenyls (PCBs), 148, 283
ppb, 11
ppm, 11
Predictive techniques, 184, 243
Prefix, 71-72
Preqard, 148-49
Prevention of contamination, 180
Profluralin, 148-49
Probe, 122-23
Progesterone, 264
Pronamide, 209, 214, 348
Propionitrile, 348
Propylbenzene, 349
S-Propyl butylethylthiocarbamate, 162-62
Propylene imine, 260
Prolate, 116
5-(2-Propenyl)-1,3-benzodioxole, 350
Propham, 149-50
Propanil, 257
Prophos, 149-50
Propuxur, 150, 281
Pyramin (e), 151
Pyrazon, 151
Pyrene, 151-52
Pyridine, 350
Pyroxychlor, 152-53

Rabon, 266
Radiolabeled chemical, 10, 285
Radon, 153, 283
Residue, 173, 292
Rf, 186, 205
Ro-Neet, 74
Ronnel, 153-54, 281
Ronstar, 140
Ruelene, 72-73
Runoff, 3, 24

Safety factor(SF), 271
Safrole, 265, 350
San 9789, 139-40
Sendran, 150
Sevin, 56
Silvex, 154, 220, 223, 282
Simazine, 155
Sodium fluoaluminate, 266
Sodium salicylkanilide, 266
Soil compartments, 5-6, 15-16
Soil contamination, 5-6
Soil environment, 173

Soil leaching, 22-24
Soil TLC, 20-21
Soil sorption, 3, 19-22
Solubility, 182
Sorption, 2, 177, 217
Spectracide, 82
Steroid hormones, 264
Steroid, 264
Styrene, 351
Sulfotep, 353
Synergistic effects, 292

2,4,5-T, 155-56
TCE, 156-57
2,3,7,8-TCDD, 352
TEA, 256
Tendex Suncide, 150
Tenoran, 70
Terbicil, 157
Terbufos, 157-58
Tersan SP, 68
Testosterone, 265
TETA, 256
Tetracene, 158
1,2,3-Tetrachlorobenzene, 354
1,2,4,5-Tetrachlorobenzene, 159
2,3,7,8-Tetrachlorodibenzo-p-dioxin, 352
1,1,1-Tetrachloroethane, 355
1,1,2,2-Tetrachloroethane, 159-60
Tetrachloroethane, 57-58, 356
1,1,1,2-Tetrachloroethane, 352
Tetrachloroethylene, 160
2,3,4,6-Tetrachlorophenol, 353
Tetraethyl dithiopyrophosphate, 291, 310, 353
Tetraethyl ester of thiodiphosphoric acid, 291, 310
Tetramethrin, 281
Thiabendazole, 161, 188, 220, 224, 258
N,N'-Thiobisphthalimide, 261
Thiocarbamate, 259, 263
Thiodemeton, 101
Tillam, 161-62
Tolban, 148-49
Toluene, 162
Toluene-2,4-diamine, 256
Toluol, 162
Torak, 79-80
Tordon, 147
Toxaphene, 162-63, 231, 236
Triallate, 163

3,4',5-Tribromosalicyanilide, 256
3,4,4-Trichlorocarbanilide, 256
Tribromomethane, 51
1,2,4-Trichlorobenzene, 164
1,1,1-Trichloroethane, 126-27
1,1,2-Trichloroethane, 355
Trichloroethylene, 156-57, 356
Trichlorofluoromethane, 164-65
Trichlorofon, 165-66
2,4,6-Trichlorophenol, 166, 282
2,4,5-Trichlorophenoxyacetic acid, 155-56, 282
1,2,3-Trichloropropane, 357
Triclopyr, 166-67
Triclopyr (butoxyethyl ester), 167-68
Triclopyr (ethylamino salt), 168
Trietazine, 168-69
Triethanolamine, 256
Triethylamine salt, 168
Triethylene tetramine, 256
Trifluralin, 169-70
1,2,4-Trimethylbenzene, 357
1,3,5-Trimethylbenzene, 358
Trolene, 153-54
Tryptophan, 261
Tunic, 122-23

Ultraviolet adsorption, 182
Urea, 170-71

Vancide 89, 55-56

Vapam, 260
Vapona, 90, 281
Vapor pressure, 18-19, 182
VC-13, 85
VCS-438, 122-23
Vinyl, 262, 263
Vinyl acetate, 358
Vinyl bromide, 262
Vinyl chloride, 171, 262, 283
Vinyl cyanide, 35-36, 262
Vinylidene chloride, 291, 311
Viozene, 153-54
Volatilization, 2, 18

Warfarin, 171-72
Water, 285
Water compartments, 5, 15
Water contamination, 5
Water solubility(WS), 9-11, 205

Xylene, 359
        ortho, 359
        meta, 359
        para, 359
m-Xylenol, 323

Zectran, 132
Zineb, 260
Zoalene, 266
Zolone, 145
Zorial, 139-40

GOVERNMENT INSTITUTES MINI-CATALOG

PC #	**ENVIRONMENTAL TITLES**	Pub Date	Price
585	Book of Lists for Regulated Hazardous Substances, 8th Edition	1997	$79
4088	CFR Chemical Lists on CD ROM, 1997 Edition	1997	$125
4089	Chemical Data for Workplace Sampling & Analysis, Single User	1997	$125
512	Clean Water Handbook, 2nd Edition	1996	$89
581	EH&S Auditing Made Easy	1997	$79
587	E H & S CFR Training Requirements, 3rd Edition	1997	$89
4082	EMMI-Envl Monitoring Methods Index for Windows-Network	1997	$537
4082	EMMI-Envl Monitoring Methods Index for Windows-Single User	1997	$179
525	Environmental Audits, 7th Edition	1996	$79
548	Environmental Engineering and Science: An Introduction	1997	$79
578	Environmental Guide to the Internet, 3rd Edition	1997	$59
560	Environmental Law Handbook, 14th Edition	1997	$79
353	Environmental Regulatory Glossary, 6th Edition	1993	$79
625	Environmental Statutes, 1998 Edition	1998	$69
4098	Environmental Statutes Book/Disk Package, 1998 Edition	1997	$208
4994	Environmental Statutes on Disk for Windows-Network	1997	$405
4994	Environmental Statutes on Disk for Windows-Single User	1997	$139
570	Environmentalism at the Crossroads	1995	$39
536	ESAs Made Easy	1996	$59
515	Industrial Environmental Management: A Practical Approach	1996	$79
4078	IRIS Database-Network	1997	$1,485
4078	IRIS Database-Single User	1997	$495
510	ISO 14000: Understanding Environmental Standards	1996	$69
551	ISO 14001: An Executive Repoert	1996	$55
518	Lead Regulation Handbook	1996	$79
478	Principles of EH&S Management	1995	$69
554	Property Rights: Understanding Government Takings	1997	$79
582	Recycling & Waste Mgmt Guide to the Internet	1997	$49
603	Superfund Manual, 6th Edition	1997	$115
566	TSCA Handbook, 3rd Edition	1997	$95
534	Wetland Mitigation: Mitigation Banking and Other Strategies	1997	$75

PC #	**SAFETY AND HEALTH TITLES**	Pub Date	Price
547	Construction Safety Handbook	1996	$79
553	Cumulative Trauma Disorders	1997	$59
559	Forklift Safety	1997	$65
539	Fundamentals of Occupational Safety & Health	1996	$49
535	Making Sense of OSHA Compliance	1997	$59
563	Managing Change for Safety and Health Professionals	1997	$59
589	Managing Fatigue in Transportation, *ATA Conference*	1997	$75
4086	OSHA Technical Manual, Electronic Edition	1997	$99
598	Project Mgmt for E H & S Professionals	1997	$59
552	Safety & Health in Agriculture, Forestry and Fisheries	1997	$125
613	Safety & Health on the Internet, 2nd Edition	1998	$49
597	Safety Is A People Business	1997	$49
463	Safety Made Easy	1995	$49
590	Your Company Safety and Health Manual	1997	$79

Electronic Product available on CD-ROM or Floppy Disk

PLEASE CALL OUR CUSTOMER SERVICE DEPARTMENT AT (301) 921-2323 FOR A FREE PUBLICATIONS CATALOG.

Government Institutes

4 Research Place, Suite 200 • Rockville, MD 20850-3226
Tel. (301) 921-2323 • FAX (301) 921-0264
E mail: giinfo@govinst.com • Internet: http://www.govinst.com

GI GOVERNMENT INSTITUTES ORDER FORM GI

4 Research Place, Suite 200 • Rockville, MD 20850-3226 • Tel (301) 921-2323 • Fax (301) 921-0264
Internet: *http://www.govinst.com* • E-mail: *giinfo@govinst.com*

3 EASY WAYS TO ORDER

1. Phone: **(301) 921-2323**
Have your credit card ready when you call.

2. Fax: **(301) 921-0264**
Fax this completed order form with your company purchase order or credit card information.

3. Mail: **Government Institutes**
4 Research Place, Suite 200
Rockville, MD 20850-3226
USA
Mail this completed order form with a check, company purchase order, or credit card information.

PAYMENT OPTIONS

❑ **Check** (*payable to Government Institutes in US dollars*)

❑ **Purchase Order** (this order form must be attached to your company P.O. Note: All International orders must be pre-paid.)

❑ **Credit Card**

Exp.___/___

Credit Card No. _____

Signature _____
Government Institutes' Federal I.D.# is 52-0994196

CUSTOMER INFORMATION

Ship To: (Please attach your Purchase Order)

Name: _____

GI Account# (*7 digits on mailing label*): _____

Company/Institution: _____

Address: _____
(please supply street address for UPS shipping)

City: _____ State/Province: _____

Zip/Postal Code: _____ Country: _____

Tel: ( ) _____

Fax: ( ) _____

E-mail Address: _____

Bill To: (if different than ship to address)

Name: _____

Title/Position: _____

Company/Institution: _____

Address: _____
(please supply street address for UPS shipping)

City: _____ State/Province: _____

Zip/Postal Code: _____ Country: _____

Tel: ( ) _____

Fax: ( ) _____

E-mail Address: _____

Qty.	Product Code	Title	Price

❑ **New Edition No Obligation Standing Order Program**

Please enroll me in this program for the products I have ordered. Government Institutes will notify me of new editions by sending me an invoice. I understand that there is no obligation to purchase the product. This invoice is simply my reminder that a new edition has been released.

15 DAY MONEY-BACK GUARANTEE

If you're not completely satisfied with any product, return it undamaged within 15 days for a full and immediate refund on the price of the product.

Subtotal_____
MD Residents add 5% Sales Tax_____
Shipping and Handling (see box below)_____
Total Payment Enclosed_____

Within U.S:	Outside U.S:
1-4 products: $6/product	Add $15 for each item (Airmail)
5 or more: $3/product	Add $10 for each item (Surface)

SOURCE CODE: BP01

Government Institutes • 4 Research Place, Suite 200 • Rockville, MD 20850
Internet: http://www.govinst.com • E-mail: giinfo@govinst.com